ZHUOMIAN YINGYONG KAIFA JICHU

桌面应用开发基础

主　编　左向荣　郝　斌　刘　群
副主编　黄利红　左国才　曾　琴
　　　　杨爱武　危孟君
主　审　符开耀　王　雷

西北工业大学出版社

【内容简介】 本书主要介绍了桌面开发的基础知识、集成开发工具的使用、使用 JDBC 或 ADO. net 操作数据库的方法和通行解决方案、基于 SQL Server2005 的数据库的创建数据库以及创建数据表、插入更新修改查询方法等内容。

本书针对具体的开发任务,阐述了解决对应开发问题的解决思路,并且提供了具体操作步骤和示范性的源代码。

本书可作为高等职业教育软件类专业的教材,也可作为桌面开发初学者的自学用书,还可供从事信息系统开发的设计、开发人员参考。

图书在版编目(CIP)数据

桌面应用开发基础/左向荣,郝斌,刘群主编. —西安:西北工业大学出版社,2016.8
ISBN 978-7-5612-4998-7

Ⅰ. ①桌… Ⅱ. ①左… ②郝… ③刘… Ⅲ. ①JAVA 语言—程序设计—高等学校—教材②C 语言—程序设计—高等学校—教材 Ⅳ. ①TP312

中国版本图书馆 CIP 数据核字(2016)第 187711 号

出版发行	:西北工业大学出版社
通信地址	:西安市友谊西路 127 号　邮编:710072
电　　话	:(029)88493844　88491757
网　　址	:www.nwpup.com
印　刷　者	:兴平市博闻印务有限公司
开　　本	:787 mm×1 092 mm　1/16
印　　张	:21.625
字　　数	:524 千字
版　　次	:2016 年 8 月第 1 版　2016 年 8 月第 1 次印刷
定　　价	:55.00 元

前　言

　　桌面开发是计算机软件开发的重要领域,有多种流行的开发技术应用于桌面开发程序设计领域,其中以基于 Java 的 SWT 技术和基于 C++的 WinForm 技术应用最为广泛。本书以 24 个真实实战项目为基础,根据学生由浅入深的习得规律,按照功能模块和岗位知识要求形成单个的开发任务。针对每个任务的完成提供必备知识、解题思路、操作步骤、具体任务及答案等内容,指导学生从一个个的实际开发任务去认识和学习桌面应用开发。

　　本书在编写风格上注重知识、技术的实用性,通过案例强化实践技能,语言力求简洁生动、通俗易懂,主要以桌面应用开发人员的岗位培养目标为核心,紧紧围绕岗位对应的职业能力和职业素质需求,选取具有典型性和代表性的项目,并以其为载体整合、序化教学内容,以实际工作任务为脉络展开教学过程,采用"项目导向、任务驱动"的方式设计课程内容的引入、示范、展开、解决、提高、实训等过程,以"教、学、做"一体化的形式带动学生自主学习。

　　在本书的编写过程中,得到了湖南软件职业学院谭长富院长、符开耀副院长、王雷教授等领导和专家们的大力支持与热心帮助,在此表示衷心感谢。

　　本书的出版还部分得到湖南软件职业学院教学质量工程项目——基于微课程模式及项目驱动模式的"C++程序设计"课程建设与研究(项目编号:KC1502)的资助;本书的部分内容参阅了国内外有关文献资料,均已经在参考文献中列出,在此一并表示感谢。

　　由于本书的编写目的定位于桌面应用开发的基础知识与案例分析相结合,试图让读者在深入了解桌面应用开发的相关概念与关键技术的基础上,能尝试开展桌面应用开发编程的一些初步编程工作,因此在本书的内容编写与结构组织上具有一定的难度,加之笔者水平有限,虽然几经修改,但书中仍然会难免存在一些疏漏与不足之处,敬请读者、专家、以及同行朋友们的批评指正,在此先行表示感谢。

　　本书凝聚了笔者多年的教学和科研经验,在编写过程中,尽管笔者一直保持严谨的态度,但难免有错误或不妥之处,恳请读者批评指正,在此深表感谢。

编　者

2016 年 6 月

目　　录

项目一　建设工程监管信息系统(一) ………………………………………………… 1
　　一、必备知识 …………………………………………………………………… 1
　　二、解题思路 …………………………………………………………………… 1
　　三、操作步骤 …………………………………………………………………… 1
　　四、具体任务 …………………………………………………………………… 1

项目二　建设工程监管信息系统(二) ………………………………………………… 5
　　一、必备知识 …………………………………………………………………… 5
　　二、解题思路 …………………………………………………………………… 5
　　三、操作步骤 …………………………………………………………………… 5
　　四、具体任务 …………………………………………………………………… 5

项目三　建设工程监管信息系统(三) ………………………………………………… 9
　　一、必备知识 …………………………………………………………………… 9
　　二、解题思路 …………………………………………………………………… 9
　　三、操作步骤 …………………………………………………………………… 9
　　四、具体任务 …………………………………………………………………… 9

项目四　码头中心船货申报系统(一) ………………………………………………… 13
　　一、必备知识 …………………………………………………………………… 13
　　二、解题思路 …………………………………………………………………… 13
　　三、操作步骤 …………………………………………………………………… 13
　　四、具体任务 …………………………………………………………………… 13

项目五　码头中心船货申报系统(二) ………………………………………………… 17
　　一、必备知识 …………………………………………………………………… 17
　　二、解题思路 …………………………………………………………………… 17
　　三、操作步骤 …………………………………………………………………… 17
　　四、具体任务 …………………………………………………………………… 17

项目六　码头中心船货申报系统(三) ………………………………………………… 21
　　一、必备知识 …………………………………………………………………… 21
　　二、解题思路 …………………………………………………………………… 21

三、操作步骤 ··· 21
四、具体任务 ··· 21

项目七 生产管理系统(一) ··· 24
一、必备知识 ··· 24
二、解题思路 ··· 24
三、操作步骤 ··· 24
四、具体任务 ··· 24

项目八 生产管理系统(二) ··· 27
一、必备知识 ··· 27
二、解题思路 ··· 27
三、操作步骤 ··· 27
四、具体任务 ··· 27

项目九 食堂饭卡管理系统 ·· 32
一、必备知识 ··· 32
二、解题思路 ··· 32
三、操作步骤 ··· 32
四、具体任务 ··· 32

项目十 建设用地供应备案系统(一) ··· 37
一、必备知识 ··· 37
二、解题思路 ··· 37
三、操作步骤 ··· 37
四、具体任务 ··· 37

项目十一 建设用地供应备案系统(二) ······································ 41
一、必备知识 ··· 41
二、解题思路 ··· 41
三、操作步骤 ··· 41
四、具体任务 ··· 41

项目十二 学生信息管理系统(一) ·· 44
一、必备知识 ··· 44
二、解题思路 ··· 44
三、操作步骤 ··· 44
四、具体任务 ··· 44

项目十三　学生信息管理系统(二) ……………………………………………………… 48
　　一、必备知识 ……………………………………………………………………………… 48
　　二、解题思路 ……………………………………………………………………………… 48
　　三、操作步骤 ……………………………………………………………………………… 48
　　四、具体任务 ……………………………………………………………………………… 48

项目十四　教务管理信息系统(一) ……………………………………………………… 51
　　一、必备知识 ……………………………………………………………………………… 51
　　二、解题思路 ……………………………………………………………………………… 51
　　三、操作步骤 ……………………………………………………………………………… 51
　　四、具体任务 ……………………………………………………………………………… 51

项目十五　教务管理信息系统(二) ……………………………………………………… 55
　　一、必备知识 ……………………………………………………………………………… 55
　　二、解题思路 ……………………………………………………………………………… 55
　　三、操作步骤 ……………………………………………………………………………… 55
　　四、具体任务 ……………………………………………………………………………… 55

项目十六　教务管理信息系统(三) ……………………………………………………… 59
　　一、必备知识 ……………………………………………………………………………… 59
　　二、解题思路 ……………………………………………………………………………… 59
　　三、操作步骤 ……………………………………………………………………………… 59
　　四、具体任务 ……………………………………………………………………………… 59

项目十七　宿舍管理系统(一) …………………………………………………………… 63
　　一、必备知识 ……………………………………………………………………………… 63
　　二、解题思路 ……………………………………………………………………………… 63
　　三、操作步骤 ……………………………………………………………………………… 63
　　四、具体任务 ……………………………………………………………………………… 63

项目十八　宿舍管理系统(二) …………………………………………………………… 68
　　一、必备知识 ……………………………………………………………………………… 68
　　二、解题思路 ……………………………………………………………………………… 68
　　三、操作步骤 ……………………………………………………………………………… 68
　　四、具体任务 ……………………………………………………………………………… 68

项目十九　宿舍管理系统(三) …………………………………………………………… 73
　　一、必备知识 ……………………………………………………………………………… 73

二、解题思路 …………………………………………………………………………… 73
　　三、操作步骤 …………………………………………………………………………… 73
　　四、具体任务 …………………………………………………………………………… 73

项目二十　通达办公自动化系统(一) …………………………………………………… 78
　　一、必备知识 …………………………………………………………………………… 78
　　二、解题思路 …………………………………………………………………………… 78
　　三、操作步骤 …………………………………………………………………………… 78
　　四、具体任务 …………………………………………………………………………… 78

项目二十一　通达办公自动化系统(二) ………………………………………………… 82
　　一、必备知识 …………………………………………………………………………… 82
　　二、解题思路 …………………………………………………………………………… 82
　　三、操作步骤 …………………………………………………………………………… 82
　　四、具体任务 …………………………………………………………………………… 82

项目二十二　通达办公自动化系统(三) ………………………………………………… 86
　　一、必备知识 …………………………………………………………………………… 86
　　二、解题思路 …………………………………………………………………………… 86
　　三、操作步骤 …………………………………………………………………………… 86
　　四、具体任务 …………………………………………………………………………… 86

项目二十三　银行信贷管理系统(一) …………………………………………………… 90
　　一、必备知识 …………………………………………………………………………… 90
　　二、解题思路 …………………………………………………………………………… 90
　　三、操作步骤 …………………………………………………………………………… 90
　　四、具体任务 …………………………………………………………………………… 90

项目二十四　银行信贷管理系统(二) …………………………………………………… 95
　　一、必备知识 …………………………………………………………………………… 95
　　二、解题思路 …………………………………………………………………………… 95
　　三、操作步骤 …………………………………………………………………………… 95
　　四、具体任务 …………………………………………………………………………… 95

附录一　Net方向部分参考答案 ………………………………………………………… 100
　　项目一 …………………………………………………………………………………… 100
　　项目二 …………………………………………………………………………………… 106
　　项目三 …………………………………………………………………………………… 112
　　项目四 …………………………………………………………………………………… 118

项目五 ……………………………………………………………………… 125
项目六 ……………………………………………………………………… 132
项目七 ……………………………………………………………………… 139
项目八 ……………………………………………………………………… 141
项目九 ……………………………………………………………………… 146
项目十 ……………………………………………………………………… 151
项目十一 …………………………………………………………………… 160
项目十二 …………………………………………………………………… 167

附录二　Java 方向部分参考答案 ………………………………………… 173
项目十三 …………………………………………………………………… 173
项目十四 …………………………………………………………………… 188
项目十五 …………………………………………………………………… 200
项目十六 …………………………………………………………………… 218
项目十七 …………………………………………………………………… 231
项目十八 …………………………………………………………………… 244
项目十九 …………………………………………………………………… 250
项目二十 …………………………………………………………………… 260
项目二十一 ………………………………………………………………… 276
项目二十二 ………………………………………………………………… 294
项目二十三 ………………………………………………………………… 312
项目二十四 ………………………………………………………………… 325

项目一 建设工程监管信息系统(一)

一、必备知识

(1)能读懂用例图,理解用户需求。
(2)能读懂类图、状态图、活动图、顺序图,理解详细设计。
(3)能使用 SWT 插件(java)或 WinForm 组件(C++)来设计窗体。
(4)数据库的设计,以及对数据库的基本操作 SQL 语句。
(5)能使用 JDBC(java)或 ADO.NET(C++)等方式建立与数据库的连接。
(6)能使用集合实现数据的存取和读出。
(7)能使用 Eclipse(java)或 Microsoft Visual Studio(C++)等开发工具并进行调试。

二、解题思路

(1)理解用例图、活动图。
(2)理解功能描述部分提供的窗体界面,使用 SWT 插件或 WinForm 设计窗体,以及窗体的布局和界面的控件。
(3)根据数据库实现提供的数据库名称和表结构,创建数据库、数据表、约束;并且在表中插入测试数据。
(4)根据功能要求,编写数据库工具类代码、界面设计及调用代码。

三、操作步骤

步骤一 创建数据库。
步骤二 界面设计。
步骤三 编写数据库工具类代码。
步骤四 编写功能或操作代码。
步骤五 按要求打包提交。

四、具体任务

1. 任务

你作为《建设工程监管信息系统》项目组的程序员,请实现下述功能:

- 用户登录；
- 工程信息查询。

2. 功能描述

(1) 如图 1.1 所示，在登录窗体中输入用户名和密码，单击"确定"按钮，进入工程查询窗体，如图 1.2 所示。

图 1.1 登录窗体

(2) 在图 1.2 中的"工程状态"下拉框中选择"正在招标"或"完成招标"，并输入"工程编号"或"工程名称"，单击"查询"按钮，将查询结果显示在"工程信息"列表中。

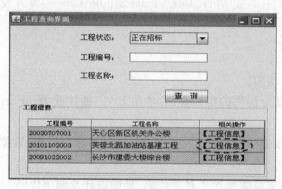

图 1.2 工程查询窗体

(3) 单击图 1.2"相关操作"列中的"工程信息"按钮，打开"工程信息"窗体，显示相应工程的详细信息，如图 1.3 所示。

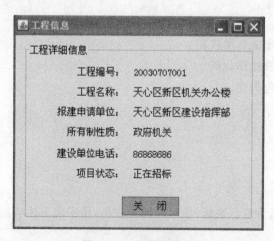

图 1.3 工程信息窗体

3. 要求

(1)窗体实现。实现图 1.1、图 1.2、图 1.3 所示窗体。

(2)数据库实现:

1)创建数据库 ConstructionDB。

2)创建管理员表(T_manager),其结构见表 1.1。

表 1.1　管理员表结构

字段名	字段说明	字段类型	是否允许为空	备注
M_id	管理员编号	Varchar(12)	否	主键
M_password	管理员密码	Varchar(12)	否	

3)在表 T_manager 中插入记录,见表 1.2。

表 1.2　T_manager 表记录

M_id	M_password
admin	admin

4)创建工程信息表(T_project),其结构见表 1.3。

表 1.3　工程信息表结构

字段名	字段说明	字段类型	是否允许为空	备注
Project_id	工程编号	Varchar(32)	否	主键
Project_name	工程名称	Varchar(64)	否	
Invi_dept	报建申请单位	Varchar(64)	否	
System_type	所有制性质	Varchar(16)	是	
Telephone	建设单位电话	Varchar(16)	是	
Project_state	项目状态	Varchar(32)	否	

5)在表 T_project 中插入记录,见表 1.4。

表 1.4　T_project 表记录

Project_id	Project_name	Invi_dept	System_type	Telephone	Project_state
20030707001	××区新区机关办公楼	××区新区建设指挥部	政府机关	86868686	正在招标
20050809002	××电子信息产业园一期工程	××电子信息产业集团有限公司	私有企业	88213462	完成招标
20101102003	××北路加油站基建工程	×××湖南分公司	国有企业	82734456	正在招标
20091022002	××市建委大楼综合楼	××市建委	政府机关	84557129	正在招标

(3)功能实现：

1)功能需求如图 1.4 所示。

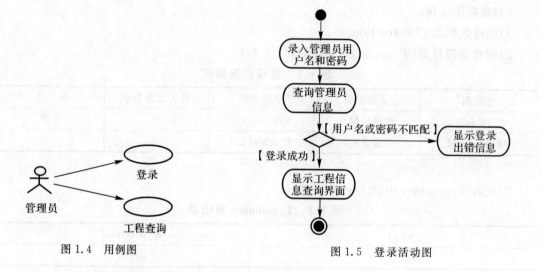

图 1.4 用例图

图 1.5 登录活动图

2)依据活动图完成管理员登录功能，如图 1.5 所示。
3)依据活动图完成查询功能，如图 1.6 所示。
工程编号和工程名称采用模糊查询，不输入工程编号和工程名称时，显示所有工程信息。

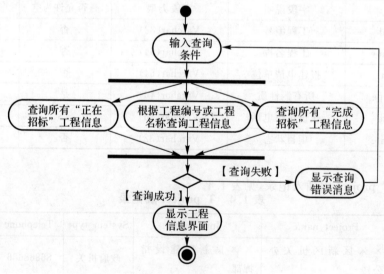

图 1.6 查询活动图

项目二 建设工程监管信息系统(二)

一、必备知识

(1)能读懂用例图,理解用户需求。
(2)能读懂类图、状态图、活动图、顺序图,理解详细设计。
(3)能使用 SWT 插件(java)或 WinForm 组件(C++)来设计窗体。
(4)数据库的设计,以及对数据库的基本操作 SQL 语句。
(5)能使用 JDBC(java)或 ADO.NET(C++)等方式建立与数据库的连接。
(6)能使用集合实现数据的存取和读出。
(7)能使用 Eclipse(java)或 Microsoft Visual Studio(C++)等开发工具并进行调试。

二、解题思路

(1)理解用例图、活动图。
(2)理解功能描述部分提供的窗体界面,使用 SWT 插件或 WinForm 设计窗体,以及窗体的布局和界面的控件。
(3)根据数据库实现提供的数据库名称和表结构,创建数据库、数据表、约束;并且在表中插入测试数据。
(4)根据功能要求,编写数据库工具类代码、界面设计及调用代码。

三、操作步骤

步骤一　创建数据库。
步骤二　界面设计。
步骤三　编写数据库工具类代码。
步骤四　编写功能或操作代码。
步骤五　按要求打包提交。

四、具体任务

1.任务

你作为《建设工程监管信息系统》项目开发组的程序员,请实现下述功能:

- 查询用户信息；
- 添加用户信息；
- 修改用户信息；
- 删除用户信息。

2. 功能描述

（1）用户信息浏览。如图 2.1 所示，单击"显示所有用户"按钮，并在左侧的"用户 ID 列表"中选择某个用户，则在右侧显示"用户姓名""用户密码"和"所属部门"。

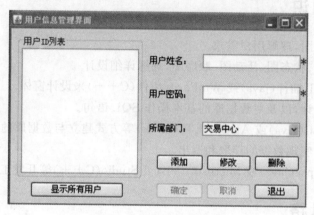

图 2.1　用户信息管理窗体

（2）添加用户。在图 2.1 中，单击"添加"按钮，输入"用户姓名""用户密码"和"所属部门"信息后，单击"确定"按钮完成用户信息添加。

（3）用户信息修改。在图 2.1 的"用户 ID 列表"中选择用户 ID，则显示相应用户信息；单击"修改"按钮，修改用户信息；单击"确定"按钮完成修改。

（4）用户信息删除。在图 2.1 的"用户 ID 列表"中选择用户 ID，单击"删除"按钮完成用户信息删除。

3. 要求

（1）界面实现。实现图 2.1 所示的用户信息管理窗体。
1)"所属部门"项值为{"交易中心""投标管理""评标委员会"}。
2)"确定"或"取消"按钮初始为不可用状态；单击"添加"或"修改"按钮，则"确定"或"取消"按钮可用；单击"确定"或"取消"按钮，则"确定"和"取消"按钮变为不可用。
3)单击"添加"按钮，则"修改"和"删除"按钮设置为不可用；单击"修改"按钮，则"添加"和"删除"按钮设置为不可用。
4)单击"取消"按钮，重新初始化用户信息管理窗体。
5)单击"退出"按钮，关闭窗体，退出应用程序。

（2）数据库实现：
1)创建数据库 ConstructionDB。
2)创建用户信息表(T_user)，其结构见表 2.1。

表 2.1　用户信息表结构

字段名	字段说明	字段类型	是否允许为空	备注
User_id	用户ID	Int	否	主键,自动增长
User_name	用户姓名	Varchar(12)	否	
User_password	用户密码	Varchar(12)	否	
Dept_name	部门名称	Varchar(32)	否	

3)在表 T_user 中插入记录,见表 2.2。

表 2.2　T_user 表记录

User_id	User_name	User_password	Dept_name
201100001	张益丰	123456	交易中心
201100002	刘伟光	676869	投标管理
201100003	李小文	888888	评标委员会
201100004	杨成武	232425	交易中心

(3)功能实现:
1)功能需求如图 2.2 所示

图 2.2　用例图

2)根据图 2.3 活动图和图 2.4 状态图完成用户添加、修改、删除和查询功能。

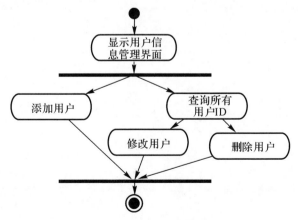

图 2.3　用户管理活动图

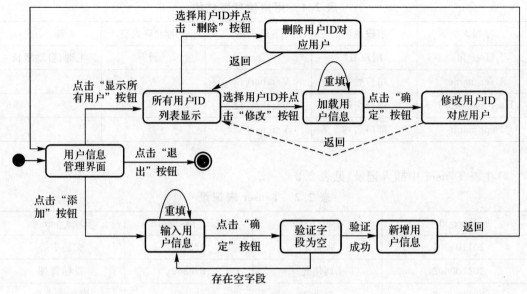

图 2.4 用户管理状态图

项目三　建设工程监管信息系统(三)

一、必备知识

(1)能读懂用例图,理解用户需求。
(2)能读懂类图、状态图、活动图、顺序图,理解详细设计。
(3)能使用 SWT 插件(java)或 WinForm 组件(C++)来设计窗体。
(4)数据库的设计,以及对数据库的基本操作 SQL 语句。
(5)能使用 JDBC(java)或 ADO.NET(C++)等方式建立与数据库的连接。
(6)能使用集合实现数据的存取和读出。
(7)能使用 Eclipse(java)或 Microsoft Visual Studio(C++)等开发工具并进行调试。

二、解题思路

(1)理解用例图、活动图。
(2)理解功能描述部分提供的窗体界面,使用 SWT 插件或 WinForm 设计窗体,以及窗体的布局和界面的控件。
(3)根据数据库实现提供的数据库名称和表结构,创建数据库、数据表、约束;并且在表中插入测试数据。
(4)根据功能要求,编写数据库工具类代码、界面设计及调用代码。

三、操作步骤

步骤一　创建数据库。
步骤二　界面设计。
步骤三　编写数据库工具类代码。
步骤四　编写功能或操作代码。
步骤五　按要求打包提交。

四、具体任务

1. 任务

你作为《建设工程监管信息系统》项目开发组的程序员,请实现下述功能:

- 查询投标企业信息；
- 新增投标信息。

2. 功能描述

(1)投标企业信息查询。在图 3.1 中,选择招标工程名称,单击"查询"按钮显示投标企业信息。

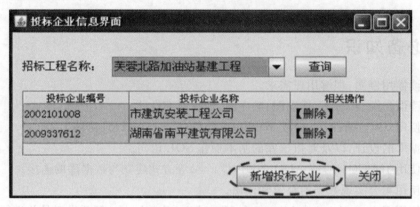

图 3.1　投标企业信息窗体

(2)新增投标信息。在图 3.1 中,单击"新增投标企业"按钮,打开图 3.2 所示窗体,选择投标企业编号,输入投标信息,单击"新增"按钮,完成投标信息添加。

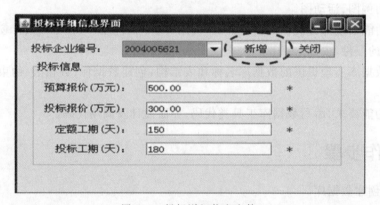

图 3.2　投标详细信息窗体

3. 要求

(1)界面实现。实现图 3.1、图 3.2 所示窗体。
1)图 3.1 中的"招标工程名称"下拉框只显示工程状态为"正在招标"的工程名称；
2)图 3.2 中的"投标企业编号"下拉框只显示投标企业编号。
(2)数据库实现：
1)创建数据库 ConstructionDB。
2)创建工程信息表(T_project),其结构见表 3.1。

表 3.1　工程信息表结构

字段名	字段说明	字段类型	是否允许为空	备注
Project_id	工程编号	Varchar(32)	否	主键
Project_name	工程名称	Varchar(64)	否	
Project_state	工程状态	Varchar(32)	否	

3) 在表 T_project 中插入记录,见表 3.2。

表 3.2　T_project 表记录

Project_id	Project_name	Project_state
20030707001	天心区新区机关办公楼	正在招标
20101102003	芙蓉北路加油站基建工程	正在招标

(4) 创建投标企业表(T_enterprise),其结构见表 3.3。

表 3.3　投标企业表结构

字段名	字段说明	字段类型	是否允许为空	备注
Ent_id	投标企业编号	Varchar(32)	否	主键
Ent_name	投标企业名称	Varchar(64)	否	

5) 在表 T_enterprise 中插入记录,见表 3.4。

表 3.4　T_enterprise 表记录

Ent_id	Ent_name
2002101008	市建筑安装工程公司
2009337612	湖南省南平建筑有限公司
2004005621	第二建筑工程有限公司

6) 创建投标信息表(T_offer),其结构见表 3.5。

表 3.5　投标信息表结构

字段名	字段说明	字段类型	是否允许为空	备注
Project_id	工程编号	Varchar(32)	否	主键,外键
Ent_id	投标企业编号	Varchar(32)	否	主键,外键
Budget_price	预算报价	Int	否	万元
Offer_price	投标报价	Int	否	万元
Ration_limite	定额工期	Int	否	天
Offer_limite	投标工期	Int	否	天

7) 在表 T_offer 中插入记录,见表 3.6。

表 3.6 T_offer 表记录

Project_id	Ent_id	Budget_price	Offer_price	Ration_limite	Offer_limite
20101102003	2002101008	500	450	150	160
20101102003	2009337612	500	480	150	200

(3)功能实现：

1)功能需求如图 3.3 所示。

2)依据活动图完成投标企业查询功能，如图 3.4 所示。

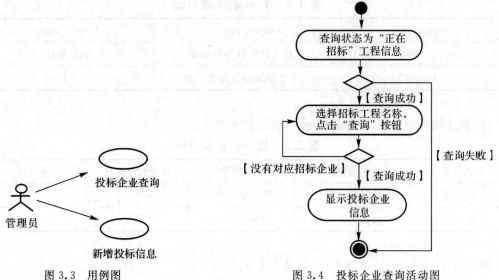

图 3.3 用例图　　　　图 3.4 投标企业查询活动图

3)依据活动图完成新增投标信息功能，如图 3.5 所示。

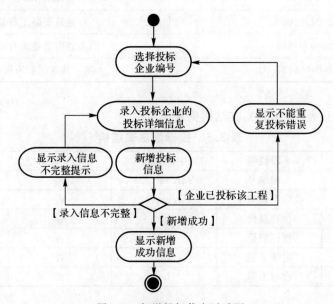

图 3.5 新增投标信息活动图

— 12 —

项目四　码头中心船货申报系统（一）

一、必备知识

(1)能读懂用例图，理解用户需求。
(2)能读懂类图、状态图、活动图、顺序图，理解详细设计。
(3)能使用 SWT 插件(java)或 WinForm 组件(C++)来设计窗体。
(4)数据库的设计，以及对数据库的基本操作 SQL 语句。
(5)能使用 JDBC(java)或 ADO.NET(C++)等方式建立与数据库的连接。
(6)能使用集合实现数据的存取和读出。
(7)能使用 Eclipse(java)或 Microsoft Visual Studio(C++)等开发工具并进行调试。

二、解题思路

(1)理解用例图、活动图。
(2)理解功能描述部分提供的窗体界面，使用 SWT 插件或 WinForm 设计窗体，以及窗体的布局和界面的控件。
(3)根据数据库实现提供的数据库名称和表结构，创建数据库、数据表、约束；并且在表中插入测试数据。
(4)根据功能要求，编写数据库工具类代码、界面设计及调用代码。

三、操作步骤

步骤一　创建数据库。
步骤二　界面设计。
步骤三　编写数据库工具类代码。
步骤四　编写功能或操作代码。
步骤五　按要求打包提交。

四、具体任务

1. 任务

你作为《码头中心船货申报系统》项目开发组的程序员，请实现下述功能：

- 查询航线信息；
- 新增航线。

2. 功能描述

(1)在图 4.1 中，选择"航线代码"、"航线名称"、"港口代码"、"码头名称"或"航线类别"等查询条件，单击"查询航线"按钮，显示查询结果，如图 4.2 所示。

图 4.1　航线管理窗体

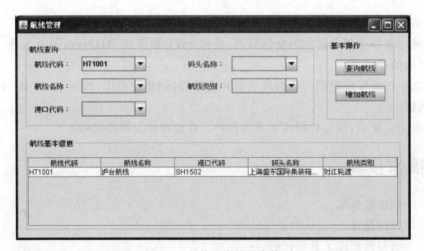

图 4.2　航线查询结果窗体

(2)在图 4.2 中单击"增加航线"按钮，打开新增航线窗体，输入"航线代码"、"航线名称"、"码头名称"和"航线类别"，选择"港口代码"，单击"保存"按钮，完成航线信息保存，如图 4.3 所示。

3. 要求

(1)窗体实现。实现图 4.1、图 4.3 所示窗体。

项目四 码头中心船货申报系统(一)

图 4.3 创健新航线界面

(2)数据库实现:

1)创建数据库 **HarborBureauDB**。

2)创建航线表(**T_line**),其结构见表 4.1。

表 4.1 航线表结构

字段名	字段说明	字段类型	是否允许为空	备注
Line_code	航线代码	**Nvarchar**(6)	否	主键
Line_name	航线名称	**Nvarchar**(50)	否	
Port_code	港口代码	**Nvarchar**(6)	否	外键
Dock_unit_name	码头名称	**Nvarchar**(50)	否	
Line_type	航线类别	**Nvarchar**(16)	否	

3)在表 **T_line** 中插入记录,见表 4.2。

表 4.2 T_line 表记录

Line_code	Line_name	Port_code	Dock_unit_name	Line_type
HT1001	沪台航线	SH1502	上海盛东国际集装箱码头有限公司	对江轮渡
TF2002	台福航线	TW1711	台湾国际港务(集团)股份有限公司	省际航线
NN4011	宁南航线	NB1563	宁波港务公司	水上游览
NS3012	南沪航线	NJ1111	南京港务(集团)股份有限公司	省际航线

4)创建港口信息表(T_port),其结构见表 4.3。

表 4.3 港口信息表结构

字段名	字段说明	字段类型	是否允许为空	备注
Port_code	港口代码	nvarchar(6)	否	主键
Port_name	港口中文名称	nvarchar(50)	否	

5)在表 T_port 插入记录,见表 4.4。

表 4.4　T_port 表记录

Port_code	Port_name
NJ1111	南京
SH1502	上海
TW1711	台湾
NB1563	宁波

(3)功能实现:

1)功能需求如图 4.4 所示。

2)依据活动图完成新增航线信息功能,如图 4.5 所示。

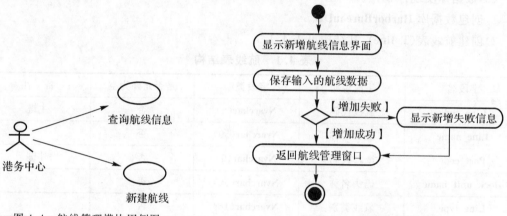

图 4.4　航线管理模块用例图　　图 4.5　新增航线活动图

3)依据活动图完成查询功能,如图 4.6 所示。

若航线代码、航线名称、港口代码、码头名称和航线类别等查询条件均为空时,显示所有航线信息;当输入一个或以上查询条件时,则显示满足所有查询条件的航线信息。

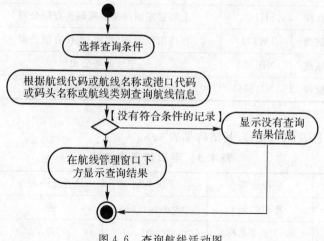

图 4.6　查询航线活动图

项目五　码头中心船货申报系统(二)

一、必备知识

(1)能读懂用例图,理解用户需求。
(2)能读懂类图、状态图、活动图、顺序图,理解详细设计。
(3)能使用 SWT 插件(java)或 WinForm 组件(C++)来设计窗体。
(4)数据库的设计,以及对数据库的基本操作 SQL 语句。
(5)能使用 JDBC(java)或 ADO.NET(C++)等方式建立与数据库的连接。
(6)能使用集合实现数据的存取和读出。
(7)能使用 Eclipse(java)或 Microsoft Visual Studio(C++)等开发工具并进行调试。

二、解题思路

(1)理解用例图、活动图。
(2)理解功能描述部分提供的窗体界面,使用 SWT 插件或 WinForm 设计窗体,以及窗体的布局和界面的控件。
(3)根据数据库实现提供的数据库名称和表结构,创建数据库、数据表、约束;并且在表中插入测试数据。
(4)根据功能要求,编写数据库工具类代码、界面设计及调用代码。

三、操作步骤

步骤一　创建数据库。
步骤二　界面设计。
步骤三　编写数据库工具类代码。
步骤四　编写功能或操作代码。
步骤五　按要求打包提交。

四、具体任务

1. 任务

你作为《码头中心船货申报系统》项目开发组的程序员,请实现下述功能:

- 用户登录;
- 修改港口设施保安员证信息。

2. 功能描述

(1)在图5.1中,输入用户名和密码,单击"确定"按钮,打开保安员证件管理窗体,显示所有港口设施保安员证件信息,如图5.2所示。

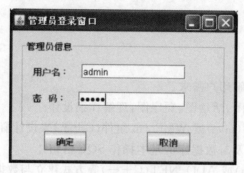

图5.1 管理员登录窗体

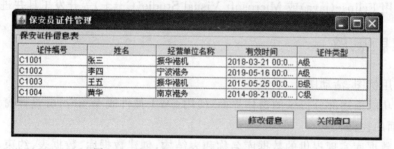

图5.2 港口设施保安员证基本信息窗体

(2)在图5.2中,选择某条记录,单击"修改信息"按钮,打开保安员证件修改窗体,如图5.3所示。其中"证件类型"的值为{"A级,B级,C级,D级"}。

(3)在图5.3中,修改保安员证件信息,单击"保存"按钮,完成保安员证件信息修改。

图5.3 港口设施保安员证基本信息修改窗体

3. 要求

(1)界面实现。实现图5.1、图5.2、图5.3所示窗体。

(2 据库实现：

1)创建数据库HarborBureauDB。

2)创建管理员表(T_manager)，其结构见表5.1。

表5.1　管理员表结构

字段名	字段说明	字段类型	是否允许为空	备　注
M_user_name	管理员用户名	Nvarchar(12)	否	主键
M_password	管理员密码	Nvarchar(12)	否	

3)在表T_manager中插入记录，见表5.2。

表5.2　T_manager表记录

M_id(管理员编号)	M_password(管理员密码)
admin	admin

4)创建保安员证件信息表(T_facility_security)，其结构见表5.3。

表5.3　保安员证件信息表结构

字段名	字段说明	字段类型	是否允许为空	备　注
Certificate_no	证书编号	Nchar(5)	否	主键
Name	姓名	Nvarchar(32)	否	
Unit_name	经营单位名称	Nvarchar(64)	否	
Effective_date	有效期	Datetime	否	
Certificate_type	证书类型	Nvarchar(32)	否	

5)在表T_facility_security中插入记录，见表5.4。

表5.4　T_facility_security表记录

Certificate_no	Name	Unit_name	Effective_date	Certificate_type
C1001	张三	振华港机	2008－3－21	A级
C1002	李四	宁波港务	2009－4－16	A级
C1003	王五	振华港机	2010－5－25	B级
C1004	黄华	南京港务	2011－8－21	C级

(3)功能实现：

1)功能需求如图5.4所示。

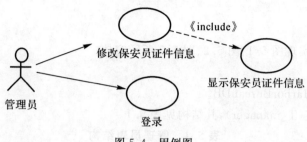

图 5.4 用例图

2）依据活动图完成管理员登录功能，如图 5.5 所示。

3）依据活动图完成修改功能，如图 5.6 所示。

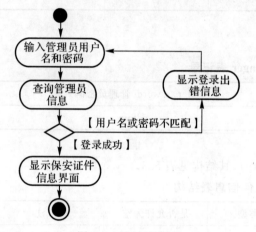

图 5.5 登录活动图

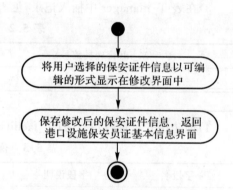

图 5.6 修改保安员证件信息活动图

项目六　码头中心船货申报系统(三)

一、必备知识

(1)能读懂用例图,理解用户需求。
(2)能读懂类图、状态图、活动图、顺序图,理解详细设计。
(3)能使用 SWT 插件(java)或 WinForm 组件(C++)来设计窗体。
(4)数据库的设计,以及对数据库的基本操作 SQL 语句。
(5)能使用 JDBC(java)或 ADO.NET(C++)等方式建立与数据库的连接。
(6)能使用集合实现数据的存取和读出。
(7)能使用 Eclipse(java)或 Microsoft Visual Studio(C++)等开发工具并进行调试。

二、解题思路

(1)理解用例图、活动图。
(2)理解功能描述部分提供的窗体界面,使用 SWT 插件或 WinForm 设计窗体,以及窗体的布局和界面的控件。
(3)根据数据库实现提供的数据库名称和表结构,创建数据库、数据表、约束;并且在表中插入测试数据。
(4)根据功能要求,编写数据库工具类代码、界面设计及调用代码。

三、操作步骤

步骤一　创建数据库。
步骤二　界面设计。
步骤三　编写数据库工具类代码。
步骤四　编写功能或操作代码。
步骤五　按要求打包提交。

四、具体任务

1.任务

你作为《码头中心船货申报系统》项目开发组的程序员,请实现下述功能:

- 显示船货信息；
- 删除指定的船货信息；
- 增加船货信息。

2. 功能描述

(1)在图 6.1 中,显示所有船货信息。

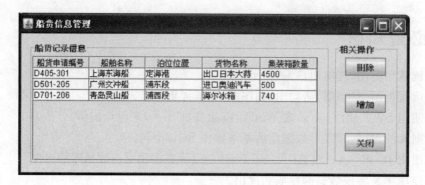

图 6.1　船货信息显示窗体

(2)在图 6.1 中,选中某条船货信息后,单击"删除"按钮弹出确认对话框,若单击"是",则删除该记录。

(3)在图 6.1 中,单击"增加"按钮,打开增加船货信息窗体,如图 6.2 所示,输入船货信息,单击"保存"按钮,完成船货信息保存。

图 6.2　创建船货记录信息界面

3. 要求

(1)界面实现。实现图 6.1、图 6.2 所示窗体。
(2)数据库实现:
1)创建数据库 HarborBureauDB。
2)创建船货信息表(T_cargo_declare),其结构见表 6.1。

表 6.1 船货信息表结构

字段名	字段说明	字段类型	是否允许为空	备注
Declare_no	船货申请编号	Nvarchar(10)	否	主键
Ship_name	船舶名称	Nvarchar(50)	否	
Berth_location	泊位位置	Nvarchar(50)	否	
Cargo_name	货物名称	Nvarchar(50)	否	
Container_qty	集装箱数量	Int	否	

3) 在表 T_cargo_declare 中插入记录,见表 6.2。

表 3.24 T_cargo_declare 表记录

Declare_no	Ship_name	Berth_location	Cargo_name	Container_qty
D501-205	广州文冲船	浦东段	进口奥迪汽车	500
D405-301	上海东海船	定海港	出口日本大蒜	4500
D701-206	青岛灵山船	浦西段	海尔冰箱	740

(3) 功能实现

1) 功能需求如图 6.3 所示。

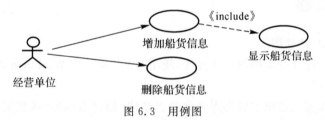

图 6.3 用例图

2) 依据活动图完成新增船货信息,并保存到 T_cargo_declare 表中,如图 6.4 所示。
3) 依据活动图完成删除功能,如图 6.5 所示。
选中被删除的记录,单击"删除"按钮,从 T_cargo_declare 表中删除该记录信息。

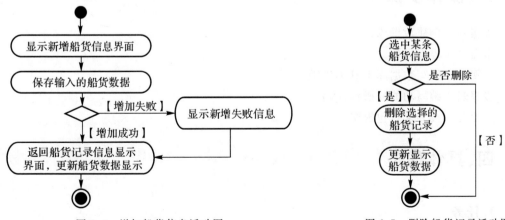

图 6.4 增加船货信息活动图　　图 6.5 删除船货记录活动图

项目七　生产管理系统（一）

一、必备知识

(1) 能读懂用例图，理解用户需求。
(2) 能读懂类图、状态图、活动图、顺序图，理解详细设计。
(3) 能使用 SWT 插件(java)或 WinForm 组件(C++)来设计窗体。
(4) 数据库的设计，以及对数据库的基本操作 SQL 语句。
(5) 能使用 JDBC(java)或 ADO.NET(C++)等方式建立与数据库的连接。
(6) 能使用集合实现数据的存取和读出。
(7) 能使用 Eclipse(java)或 Microsoft Visual Studio(C++)等开发工具并进行调试。

二、解题思路

(1) 理解用例图、活动图。
(2) 理解功能描述部分提供的窗体界面，使用 SWT 插件或 WinForm 设计窗体，以及窗体的布局和界面的控件。
(3) 根据数据库实现提供的数据库名称和表结构，创建数据库、数据表、约束；并且在表中插入测试数据。
(4) 根据功能要求，编写数据库工具类代码、界面设计及调用代码。

三、操作步骤

步骤一　创建数据库。
步骤二　界面设计。
步骤三　编写数据库工具类代码。
步骤四　编写功能或操作代码。
步骤五　按要求打包提交。

四、具体任务

1. 任务

你作为《生产管理系统》项目开发组的程序员，请实现下述功能：

- 按类别查询产品信息。

2. 功能描述

(1)在图 7.1 中，显示所有产品信息。

(2)在图 7.1 中，选择"产品类别"，单击"查询"按钮，显示该类别的产品信息，如图 7.2 所示；单击"全部产品"，显示所有产品信息。

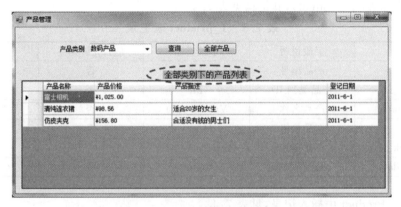

图 7.1　显示全部类别的产品信息

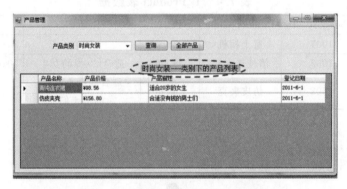

图 7.2　显示某个类别下的产品列表

3. 要求

(1)界面实现。实现图 7.1、图 7.2 所示窗体。

(2)数据库实现：

1)创建数据库 ProductDB。

2)创建产品类别表(T_category)，其结构见表 7.1。

表 7.1　产品类别表结构

字段名	字段说明	字段类型	是否允许为空	备注
Category_id	产品类别 ID	Int	否	主键
Category_name	产品类别名称	Varchar(30)	否	
Register_date	产品类别产生时间	Datetime	否	

3)在表 T_category 中插入记录,见表 7.2。

表 7.2　T_category 表记录

Category_id	Category_name	Register_date
1001	数码产品	2011-6-1 14:34:45
1002	时尚女装	2010-6-1 09:30:15

4)创建产品表(T_product),其结构见表 7.3。

表 7.3　T_product 表记录

字段名	字段说明	字段类型	是否允许为空	备注
Product_id	产品编号	Int	否	主键
Category_id	产品类别 ID	Int	否	外键
Product_name	产品名称	Varchar(50)	否	
Price	产品价格	Money	否	
Remark	产品描述	Nvarchar(2000)		
Register_date	产品录入时间	Datetime	否	

5)在表 T_Product 中插入记录,见表 7.4。

表 7.4　T_Product 表数据

Product_id	Category_id	Product_name	Price	Remark	Register_date
20110001	1001	富士相机	1025.00		2011-6-1 14:34:45
20110002	1002	清纯连衣裙	98.56	适合 20 岁的女生	2010-6-1 09:30:15
20110003	1002	仿皮夹克	156.80	适合经济不宽松的男士们	2010-6-1 15:35:00

(3)功能实现:

1)依据活动图完成产品信息查询,如图 7.3 所示。

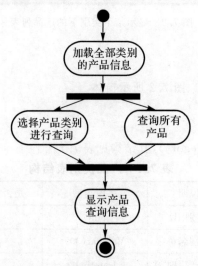

图 7.3　按类别查询产品活动图

项目八　生产管理系统(二)

一、必备知识

(1)能读懂用例图,理解用户需求。
(2)能读懂类图、状态图、活动图、顺序图,理解详细设计。
(3)能使用 SWT 插件(java)或 WinForm 组件(C++)来设计窗体。
(4)数据库的设计,以及对数据库的基本操作 SQL 语句。
(5)能使用 JDBC(java)或 ADO.NET(C++)等方式建立与数据库的连接。
(6)能使用集合实现数据的存取和读出。
(7)能使用 Eclipse(java)或 Microsoft Visual Studio(C++)等开发工具并进行调试。

二、解题思路

(1)理解用例图、活动图。
(2)理解功能描述部分提供的窗体界面,使用 SWT 插件或 WinForm 设计窗体,以及窗体的布局和界面的控件。
(3)根据数据库实现提供的数据库名称和表结构,创建数据库、数据表、约束;并且在表中插入测试数据。
(4)根据功能要求,编写数据库工具类代码、界面设计及调用代码。

三、操作步骤

步骤一　创建数据库。
步骤二　界面设计。
步骤三　编写数据库工具类代码。
步骤四　编写功能或操作代码。
步骤五　按要求打包提交。

四、具体任务

1. 任务

你作为《生产管理系统》项目开发组的程序员,请实现下述功能:

- 浏览产品信息；
- 添加产品信息。

2. 功能描述

(1)在图 8.1 中，显示所有产品信息。

(2)在图 8.2 中，输入新增产品信息，单击"添加产品"按钮，数据校验通过后，保存产品信息，并显示"已经成功添加数据"对话框，如图 8.3 所示，单击"确定"，清除文本框所输入的内容，如图 8.4 所示。

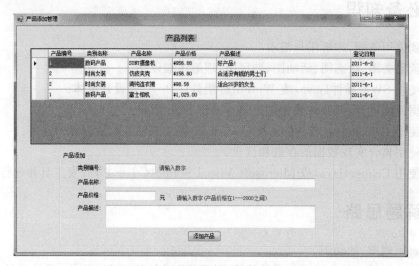

图 8.1 产品信息显示及产品添加界面

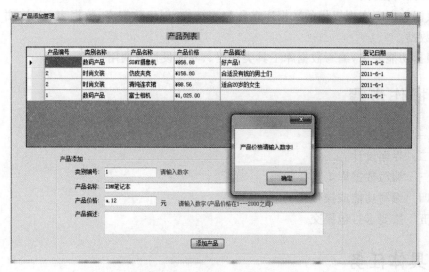

图 8.2 添加时检查数据合法性

图 8.3 添加成功之后,将新添加的产品显示在表格中

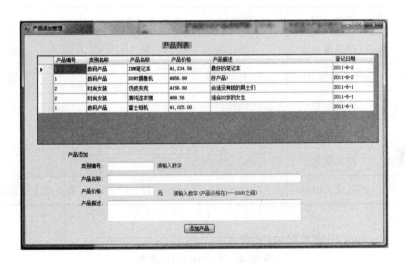

图 8.4 用户单击对话框的"确定"后,清除文本框所输入内容

3. 要求

(1)界面实现

实现图 8.1～图 8.4 所示界面,其中在图 8.1 中,提示产品添加信息有效性的标签,要求用红色显示。

(2)数据库实现:

1)创建数据库 ProductDB。

2)创建产品类别表(T_category),其结构见表 8.1。

表 8.1　产品类别表结构

字段名	字段说明	字段类型	是否允许为空	备注
Category_id	产品类别 ID	Int	否	主键
Category_name	产品类别名称	Varchar(30)	否	
Register_date	产品类别产生时间	Datetime	否	默认值为当前录入时间

3)在表 T_category 中插入记录,见表 8.2。

表 8.2　T_category 表记录

Category_id	Category_name	Register_date
1001	数码产品	2011-6-1 14:34:45
1002	时尚女装	2010-6-1 09:30:15

4)创建产品表(T_Product),其结构见表 8.3。

表 8.3　产品表结构

字段名	字段说明	字段类型	是否允许为空	备注
Product_id	产品编号	Int	否	主键
Category_id	产品类别 ID	Int	否	外键
Product_name	产品名称	Varchar(50)	否	
Price	产品价格	Money	否	
Remark	产品描述	Nvarchar(2000)		
Register_date	产品录入时间	Datetime	否	默认值为当前录入时间

5)在表 T_Product 中插入记录,见表 8.4。

表 8.4　T_Product 记录

Product_id	Category_id	Product_name	Price	Remark	Register_date
20110001	1001	富士相机	1025.00		2011-6-1 14:34:45
20110002	1002	清纯连衣裙	98.56	适合 20 岁的女生	2010-6-1 09:30:15
20110003	1002	仿皮夹克	156.80	适合经济不宽松的男士们	2010-6-1 15:35:00

(3)功能实现。依据活动图完成产品信息增加,如图 8.5 所示。

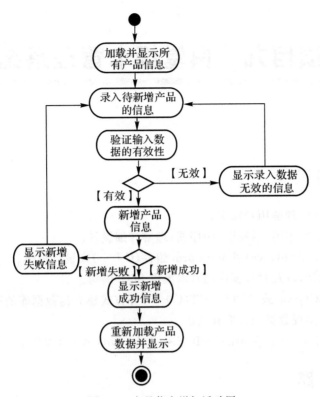

图 8.5 产品信息增加活动图

项目九 食堂饭卡管理系统

一、必备知识

(1)能读懂用例图,理解用户需求。
(2)能读懂类图、状态图、活动图、顺序图,理解详细设计。
(3)能使用 SWT 插件(java)或 WinForm 组件(C++)来设计窗体。
(4)数据库的设计,以及对数据库的基本操作 SQL 语句。
(5)能使用 JDBC(java)或 ADO.NET(C++)等方式建立与数据库的连接。
(6)能使用集合实现数据的存取和读出。
(7)能使用 Eclipse(java)或 Microsoft Visual Studio(C++)等开发工具并进行调试。

二、解题思路

(1)理解用例图、活动图。
(2)理解功能描述部分提供的窗体界面,使用 SWT 插件或 WinForm 设计窗体,以及窗体的布局和界面的控件。
(3)根据数据库实现提供的数据库名称和表结构,创建数据库、数据表、约束;并且在表中插入测试数据。
(4)根据功能要求,编写数据库工具类代码、界面设计及调用代码。

三、操作步骤

步骤一　创建数据库。
步骤二　界面设计。
步骤三　编写数据库工具类代码。
步骤四　编写功能或操作代码。
步骤五　按要求打包提交。

四、具体任务

1. 任务

你作为《食堂饭卡管理系统》项目开发组的程序员,请实现下述功能:

- 浏览饭卡信息；
- 添加饭卡信息。

2. 功能描述

(1)在图 9.1 中,显示所有饭卡信息。

(2)在图 9.2 中,输入新增饭卡信息,单击"添加饭卡信息"按钮,数据校验通过后,保存饭卡信息,并显示"恭喜你添加成功"对话框,如图 9.3 所示,单击"确定",清除文本框所输入的内容,如图 9.4 所示。

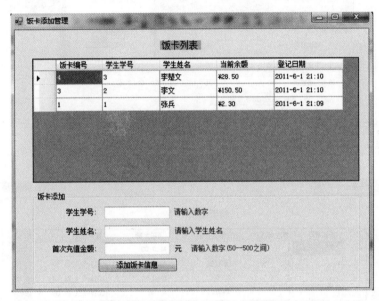

图 9.1 饭卡信息显示及添加功能

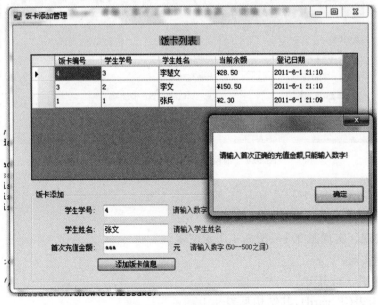

图 9.2 检查数据合法性

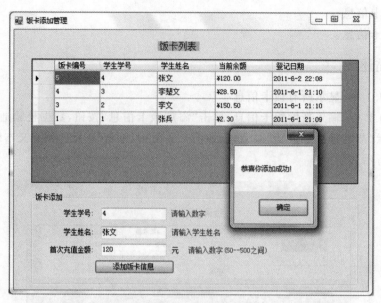

图 9.3 添加成功后,将新添加的饭卡信息显示在界面表格中

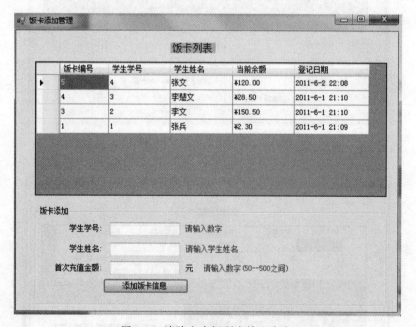

图 9.4 清除文本框所有输入内容

3. 要求

(1)界面实现。实现图 9.1～图 9.4 所示窗体。

(2)数据库实现:

1)创建数据库 CardDB。

2)创建饭卡表(T_card),其结构见表 9.1。

表 9.1 饭卡表结构

字段名	字段说明	字段类型	是否允许为空	备注
Card_id	饭卡编号	Int	否	主键
Student_id	学生学号	Int	否	
Student_name	学生姓名	Varchar(10)	否	
Curr_money	饭卡余额（初始值为 0）	Money		
Register_date	饭卡生成时间	Datetime	否	默认值为当前录入时间

3）在表 T_card 中插入记录，见表 9.2。

表 9.2 T_card 表记录

Card_id	Student_id	Student_name	Curr_money	Register_date
1	1	李楚文	28.50	2011-6-1 21:10:45
3	2	李文	150.50	2010-6-1 21:10:15
4	3	张兵	2.30	2010-6-1 21:09:00

4）创建饭卡充值表（T_add_money），表结构见表 9.3。

表 9.3 饭卡充值表结构

字段名	字段说明	字段类型	是否允许为空	备注
Add_id	充值编号	Int	否	主键
Card_id	饭卡编号	Int	否	外键
The_money	充值金额	Money	否	充值金额为 50 元到 200 元之间
Register_date	充值时间，默认值为当前录入时间	Datetime	否	默认值为当前录入时间

5）在表 T_add_money 中插入记录，见表 9.4。

表 9.4 T_add_money 表记录

Add_id	Card_id	The_money	Register_date
1	1	50.00	2011-6-6 14:34:45
2	2	100.00	2010-6-8 09:30:15
3	3	180.00	2010-6-2 15:35:00

（3）功能实现。依据活动图完成饭卡信息增加，如图 9.5 所示。

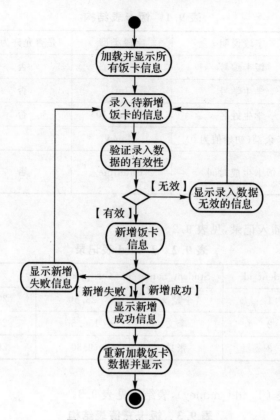

图 9.5 饭卡信息增加活动图

项目十 建设用地供应备案系统(一)

一、必备知识

(1)能读懂用例图,理解用户需求。
(2)能读懂类图、状态图、活动图、顺序图,理解详细设计。
(3)能使用 SWT 插件(java)或 WinForm 组件(C++)来设计窗体。
(4)数据库的设计,以及对数据库的基本操作 SQL 语句。
(5)能使用 JDBC(java)或 ADO.NET(C++)等方式建立与数据库的连接。
(6)能使用集合实现数据的存取和读出。
(7)能使用 Eclipse(java)或 Microsoft Visual Studio(C++)等开发工具并进行调试。

二、解题思路

(1)理解用例图、活动图。
(2)理解功能描述部分提供的窗体界面,使用 SWT 插件或 WinForm 设计窗体,以及窗体的布局和界面的控件。
(3)根据数据库实现提供的数据库名称和表结构,创建数据库、数据表、约束;并且在表中插入测试数据。
(4)根据功能要求,编写数据库工具类代码、界面设计及调用代码。

三、操作步骤

步骤一　创建数据库。
步骤二　界面设计。
步骤三　编写数据库工具类代码。
步骤四　编写功能或操作代码。
步骤五　按要求打包提交。

四、具体任务

1. 任务

你作为《建设用地供应备案系统》项目开发组的程序员,请实现下述功能:

- 项目查询；
- 增加新项目。

2. 功能描述

(1)在图10.1中，单击"查询"菜单，选择项目名称，显示项目详细信息，如图10.2所示。

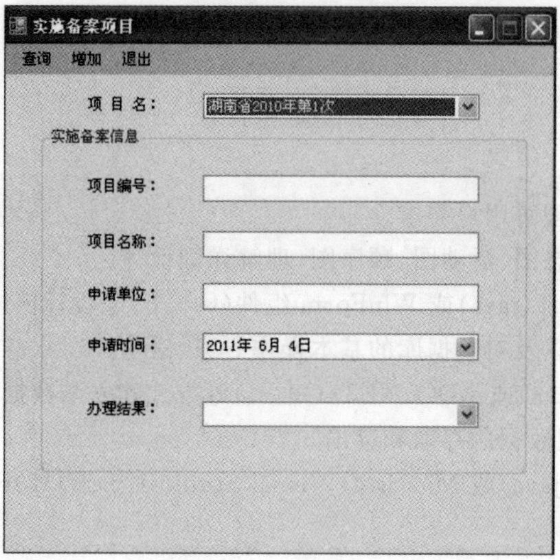

图 10.1　查询窗体

图 10.2　详细信息窗体

(2)在图10.1中，单击"增加"菜单，窗体显示如图10.3所示，输入新增项目信息，单击"确认"按钮，若显示图10.4所示对话框，则新增项目信息保存成功，若显示图10.5所示对话框，则新增项目信息失败。

— 38 —

项目十　建设用地供应备案系统(一)

图 10.3　增加信息窗体

图 10.4　添加成功窗体

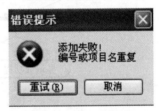

图 10.5　添加失败窗体

3. 要求

(1)界面实现。实现图 10.1～图 10.2 所示窗体。

(2)数据库实现：

1)创建数据库 SupplyDB。

2)创建项目信息表(T_record)，其结构见表 10.1。

表 10.1　项目信息表结构

字段名	字段说明	字段类型	是否允许为空	备注
Guid	项目编号	Varchar(10)	否	主键
Proj_name	项目名称	Varchar(50)	否	唯一键
Make_unit	申请单位	Varchar(50)	否	
Time	申请时间	Datetime	否	
Result	处理结果	Varchar(10)	否	

3)在表 T_record 中插入记录，见表 10.2。

表 10.2 T_record 表记录

Guid(项目编号)	Proj_name(项目名称)	Make_unit(申请单位)	Time(申请时间)	Result(处理结果)
zj20110604	湖南省 2010 年第 1 次	联想	2010-6-2	成功
zj20110605	湖南省 2011 年第 1 次	湖南省人民政府	2011-2-2	待审

(3)功能实现：

1)依据活动图完成项目增加功能,如图 10.6 所示。

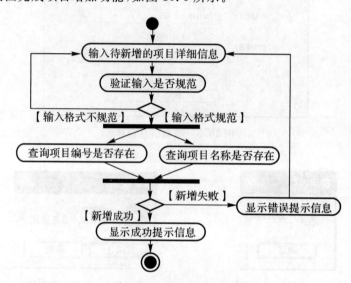

图 10.6　增加项目活动图

2)依据活动图完成项目查询功能,如图 10.7 所示。

图 10.7　项目查询活动图

项目十一　建设用地供应备案系统(二)

一、必备知识

(1)能读懂用例图,理解用户需求。
(2)能读懂类图、状态图、活动图、顺序图,理解详细设计。
(3)能使用 SWT 插件(java)或 WinForm 组件(C++)来设计窗体。
(4)数据库的设计,以及对数据库的基本操作 SQL 语句。
(5)能使用 JDBC(java)或 ADO.NET(C++)等方式建立与数据库的连接。
(6)能使用集合实现数据的存取和读出。
(7)能使用 Eclipse(java)或 Microsoft Visual Studio(C++)等开发工具并进行调试。

二、解题思路

(1)理解用例图、活动图。
(2)理解功能描述部分提供的窗体界面,使用 SWT 插件或 WinForm 设计窗体,以及窗体的布局和界面的控件。
(3)根据数据库实现提供的数据库名称和表结构,创建数据库、数据表、约束;并且在表中插入测试数据。
(4)根据功能要求,编写数据库工具类代码、界面设计及调用代码。

三、操作步骤

步骤一　创建数据库。
步骤二　界面设计。
步骤三　编写数据库工具类代码。
步骤四　编写功能或操作代码。
步骤五　按要求打包提交。

四、具体任务

1. 任务

你作为《建设用地供应备案系统》项目开发组的程序员,请实现下述功能:

- 查询所有勘测定界资料；
- 修改勘测定界资料。

2. 功能描述

(1)在图 11.1 中，显示所有勘测定界资料信息。

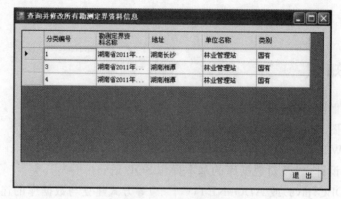

图 11.1　查询窗体

(2)在图 11.2 中，选择某条记录，右键单击，在弹出菜单中单击"修改"菜单，编辑勘测定界信息，修改完成后在弹出菜单中单击"保存"菜单，完成新增信息保存，并显示"修改成功"对话框，如图 11.3 所示。

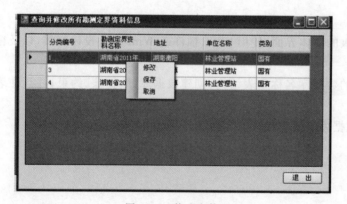

图 11.2　修改窗体

图 11.3　修改成功对话框

3. 要求

(1)界面实现。实现图 11.1、图 11.2 所示窗体。

(2)数据库实现：

1)创建数据库 QualificationDB。

2)创建勘测定界信息表(T_qualification)，其结构见表 11.1。

表 11.1 勘测定界信息表结构

字段名	字段说明	字段类型	是否允许为空	备注
Ca_guid	土地分类编号	Int	否	主键
Sb_name	勘测定界资料名称	Varchar(20)	否	
Address	地址	Varchar(50)	否	
Unit_name	单位名称	Varchar(50)	否	
The_owner	国有或集体	Varchar(30)	否	

3)在表 T_qualification 中插入记录,见表 11.2。

表 11.2 T_qualification 表记录

Ca_guid	Sb_name	Address	Unit_name	The_owner
1	××省2011年第二次	×××市	林业管理局	国有
3	××省2011年第二次	×××市	林业管理局	国有
4	××省2011年第一次	×××市	林业管理局	国有

(3)功能实现

1)依据活动图完成所有勘测定界资料查询功能,如图 11.4 所示。

2)依据活动图完成勘测定界资料修改功能,如图 11.5 所示。

图 11.4 查询活动图

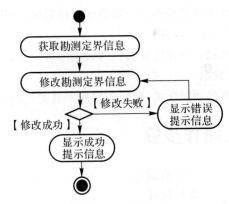

图 11.5 修改活动图

项目十二　学生信息管理系统(一)

一、必备知识

(1)能读懂用例图,理解用户需求。
(2)能读懂类图、状态图、活动图、顺序图,理解详细设计。
(3)能使用 SWT 插件(java)或 WinForm 组件(C++)来设计窗体。
(4)数据库的设计,以及对数据库的基本操作 SQL 语句。
(5)能使用 JDBC(java)或 ADO.NET(C++)等方式建立与数据库的连接。
(6)能使用集合实现数据的存取和读出。
(7)能使用 Eclipse(java)或 Microsoft Visual Studio(C++)等开发工具并进行调试。

二、解题思路

(1)理解用例图、活动图。
(2)理解功能描述部分提供的窗体界面,使用 SWT 插件或 WinForm 设计窗体,以及窗体的布局和界面的控件。
(3)根据数据库实现提供的数据库名称和表结构,创建数据库、数据表、约束;并且在表中插入测试数据。
(4)根据功能要求,编写数据库工具类代码、界面设计及调用代码。

三、操作步骤

步骤一　创建数据库。
步骤二　界面设计。
步骤三　编写数据库工具类代码。
步骤四　编写功能或操作代码。
步骤五　按要求打包提交。

四、具体任务

1. 任务

你作为《学生信息管理系统》项目开发组的程序员,请实现下述功能:

- 用户登录；
- 查询学生信息。

2. 功能描述

(1)在图12.1中，输入用户名和密码，单击"登录"按钮，打开学生信息查询窗体，如图12.1所示。

图 12.1 登录窗体

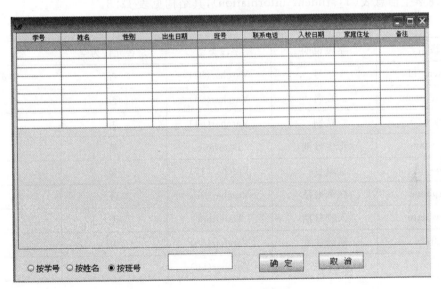

图 12.2 学生信息查询窗体

(2)在图12.2中，输入学生学号或姓名或班级号，单击"确定按钮"，显示查询结果。

3. 要求

(1)界面实现。实现图12.1、图12.2所示窗体。

(2)数据库实现：

1)创建数据库 studentDB。

2)创建用户表(T_user)，其结构见表12.1。

表 12.1 用户表结构

字段名	字段说明	字段类型	是否允许为空	备注
User_id	编号	Varchar(12)	否	主键
User_name	用户名	Varchar(50)	否	
User_password	密码	Varchar(12)	否	

3)在表 T_user 中插入记录,见表 12.2。

表 12.2 T_user 表记录

User_id	User_name	User_password
admin	admin	123456
hehe	hehe	111111

4)创建学生信息表(T_student_information),其结构见表 12.3。

表 12.3 学生信息表表结构

字段名	字段说明	字段类型	是否允许为空	备注
Student_id	学号	Varchar(32)	否	主键
Student_name	姓名	Varchar(64)	否	
Sex	性别	Varchar(64)	否	
Birthday	出生日期	Datetime	是	
Class_id	班号	Varchar(16)	否	
Telephone	联系电话	Varchar(32)	否	
Entry_date	入校日期	Datetime	否	
Address	家庭住址	Varchar(50)	否	
Memo	备注	Varchar(50)	是	

5)在表 T_student_information 中插入记录,见表 12.4。

表 12.4 T_student_information 表记录

Student_id	Student_name	Sex	Birthday	Class_id	Telephone	Entry_date	Address	Memo
1001	张三	f	1993.3.8	102	56726	2010.9.5	××市	
1002	刘青	m	1993.5.1	101	23456	2010.9.5	××市	
1003	李艳	f	1992.4.9	102	43567	2010.9.5	××市	
1004	王杰	f	1995.1.3	101	23456	2010.9.5	××市	

(3)功能实现

1)功能需求如图12.3所示。
2)依据活动图完成用户登录功能,如图12.4所示。

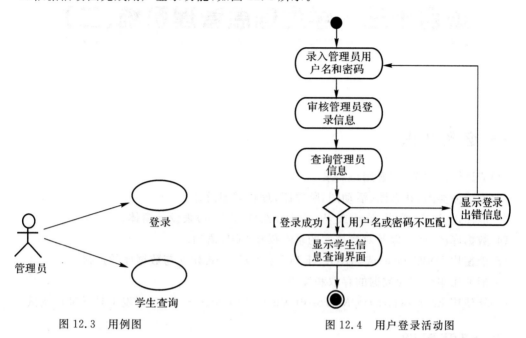

图12.3 用例图　　　　图12.4 用户登录活动图

3)依据活动图完成学生查询功能,如图12.5所示。

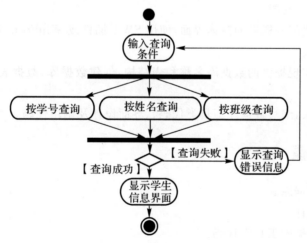

图12.5 学生查询活动图

项目十三　学生信息管理系统(二)

一、必备知识

(1)能读懂用例图,理解用户需求。
(2)能读懂类图、状态图、活动图、顺序图,理解详细设计。
(3)能使用 SWT 插件(java)或 WinForm 组件(C++)来设计窗体。
(4)数据库的设计,以及对数据库的基本操作 SQL 语句。
(5)能使用 JDBC(java)或 ADO.NET(C++)等方式建立与数据库的连接。
(6)能使用集合实现数据的存取和读出。
(7)能使用 Eclipse(java)或 Microsoft Visual Studio(C++)等开发工具并进行调试。

二、解题思路

(1)理解用例图、活动图。
(2)理解功能描述部分提供的窗体界面,使用 SWT 插件或 WinForm 设计窗体,以及窗体的布局和界面的控件。
(3)根据数据库实现提供的数据库名称和表结构,创建数据库、数据表、约束;并且在表中插入测试数据。
(4)根据功能要求,编写数据库工具类代码、界面设计及调用代码。

三、操作步骤

步骤一　创建数据库。
步骤二　界面设计。
步骤三　编写数据库工具类代码。
步骤四　编写功能或操作代码。
步骤五　按要求打包提交。

四、具体任务

1. 任务

你作为《学生信息管理系统》项目开发组的程序员,请实现下述功能:

·修改用户密码。

2. 功能描述

(1)在图 13.1 中,单击"系统"→"修改密码"菜单,打开修改密码窗体,如图 13.2 所示。

图 13.1　主界面

(2)在图 13.2 中,输入用户名和密码,单击"确定"按钮,显示如图 13.3 所示对话框框,单击对话框"确认"按钮,完成密码修改。

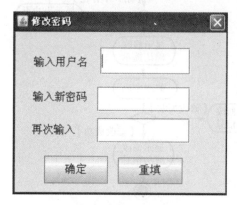

图 13.2　修改密码窗体

图 13.3　修改密码确认对话框

3. 要求

(1)界面实现。实现图 13.1、图 13.2 所示窗体。
(2)数据库实现:
1)创建数据库 studentDB。

2)创建用户表(T_user),其结构见表 13.1。

表 13.1 用户表结构

字段名	字段说明	字段类型	是否允许为空	备 注
User_id	编号	Varchar(12)	否	主键
User_name	用户名	Varchar(50)	否	
User_password	密码	Varchar(12)	否	

3)在表 T_user 中插入记录,见表 13.2。

表 13.2 T_user 表记录

User_id	User_name	User_password
admin	admin	123456
hehe	hehe	111111

(3)功能实现:

1)功能需求如图 13.4 所示。

2)依据活动图完成用户密码修改功能,如图 13.5 所示。

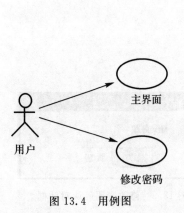

图 13.4 用例图

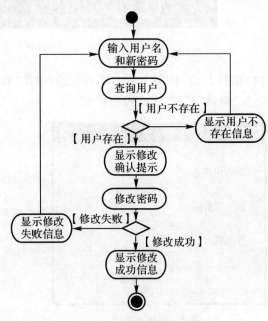

图 13.5 修改密码活动图

项目十四　教务管理信息系统（一）

一、必备知识

（1）能读懂用例图，理解用户需求。
（2）能读懂类图、状态图、活动图、顺序图，理解详细设计。
（3）能使用 SWT 插件(java)或 WinForm 组件(C++)来设计窗体。
（4）数据库的设计，以及对数据库的基本操作 SQL 语句。
（5）能使用 JDBC(java)或 ADO.NET(C++)等方式建立与数据库的连接。
（6）能使用集合实现数据的存取和读出。
（7）能使用 Eclipse(java)或 Microsoft Visual Studio(C++)等开发工具并进行调试。

二、解题思路

（1）理解用例图、活动图。
（2）理解功能描述部分提供的窗体界面，使用 SWT 插件或 WinForm 设计窗体，以及窗体的布局和界面的控件。
（3）根据数据库实现提供的数据库名称和表结构，创建数据库、数据表、约束；并且在表中插入测试数据。
（4）根据功能要求，编写数据库工具类代码、界面设计及调用代码。

三、操作步骤

步骤一　创建数据库。
步骤二　界面设计。
步骤三　编写数据库工具类代码。
步骤四　编写功能或操作代码。
步骤五　按要求打包提交。

四、具体任务

1. 任务

你作为《教务管理信息系统》项目开发组的程序员，请实现下述功能：

- 添加供应商；
- 查询订单信息。

2.功能描述

(1)在图 14.1 中,单击"添加供应商信息"选项卡,输入供应商信息,单击"添加"按钮,完成供应商信息保存。

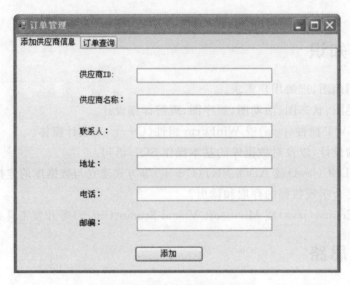

图 14.1　供应商信息添加界面

(2)在图 14.1 中,单击"订单查询"选项卡,显示如图 14.2 所示窗体,输入订单编号,单击"查询"按钮,显示查询结果;若未输入订单编号,则显示所有订单信息。

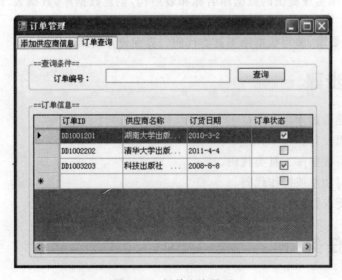

图 14.2　订单查询界面

3. 要求

(1)界面实现。实现图 14.1、图 14.2 所示窗体。

(2)数据库实现：

1)创建数据库 HNIUEAMDB。

2)创建供应商信息表(T_supplier_information)，其结构见表 14.1。

表 14.1　供应商信息表结构

字段名	字段说明	字段类型	是否允许为空	备注
Supplier_id	供应商编号	Varchar(10)	否	主键
Supplier_name	供应商名称	Varchar(50)	否	
Supplier_people	供应商联系人	Varchar(8)	否	
Supplier_address	供应商地址	Varchar(50)	是	
Supplier_phone	供应商电话	Varchar(11)	是	
Supplier_code	供应商邮编	Varchar(6)	是	

3)在表 T_supplier_information 中插入记录，见表 14.2。

表 14.2　T_supplier_information 表记录

Supplier_id	Supplier_name	Supplier_people	Supplier_address	Supplier_phone	Supplier_code
BJ1002	清华大学出版社	郭政强	北京清华大学	15123467890	101023
CD1003	科技出版社	蒋军	成都电子科技大学	15874679856	290897
CS1001	湖南大学出版社	李伟	湖南大学	13789654673	410230

4)创建订单表(T_order)，其结构见表 14.3。

表 14.3　订单表表结构

字段名	字段说明	字段类型	是否允许为空	备注
Order_id	订单编号	Varchar(10)	否	主键
Supplier_id	供应商 ID	Varchar(10)	否	
Order_date	订货日期	Datetime	否	
Order_status	订单状态	Bit	否	

5)在表 T_order 中插入记录，见表 14.4。

表 14.4　T_order 表记录

Order_id	Supplier_id	Order_date	Order_status
DD1001201	CS1001201003	2010-03-02 00:00:00.000	True
DD1002202	BJ1002201104	2011-04-04 00:00:00.000	False
DD1003203	CD1003200808	2008-08-08 00:00:00.000	True

(3)功能实现:

1)功能需求如图 14.3 所示。

2)依据活动图完成供应商信息添加功能,如图 14.4 所示。

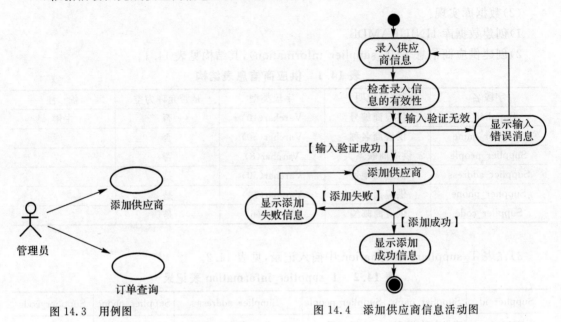

图 14.3 用例图

图 14.4 添加供应商信息活动图

3)依据活动图完成查询功能,如图 14.5 所示。

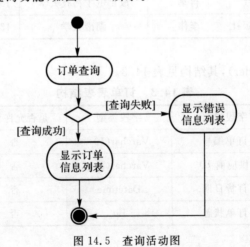

图 14.5 查询活动图

项目十五　教务管理信息系统(二)

一、必备知识

(1)能读懂用例图,理解用户需求。
(2)能读懂类图、状态图、活动图、顺序图,理解详细设计。
(3)能使用 SWT 插件(java)或 WinForm 组件(C++)来设计窗体。
(4)数据库的设计,以及对数据库的基本操作 SQL 语句。
(5)能使用 JDBC(java)或 ADO.NET(C++)等方式建立与数据库的连接。
(6)能使用集合实现数据的存取和读出。
(7)能使用 Eclipse(java)或 Microsoft Visual Studio(C++)等开发工具并进行调试。

二、解题思路

(1)理解用例图、活动图。
(2)理解功能描述部分提供的窗体界面,使用 SWT 插件或 WinForm 设计窗体,以及窗体的布局和界面的控件。
(3)根据数据库实现提供的数据库名称和表结构,创建数据库、数据表、约束;并且在表中插入测试数据。
(4)根据功能要求,编写数据库工具类代码、界面设计及调用代码。

三、操作步骤

步骤一　创建数据库。
步骤二　界面设计。
步骤三　编写数据库工具类代码。
步骤四　编写功能或操作代码。
步骤五　按要求打包提交。

四、具体任务

1. 任务

你作为《教务管理信息系统》项目开发组的程序员,请实现下述功能:

- 查询教材信息；
- 修改教材信息。

2. 功能描述

(1)在图 15.1 中，输入教材名称，单击"查询"按钮，将在教材信息列表中显示查询结果。任选一条教材信息，并单击"修改"按钮，弹出如图 15.2 所示窗体。

图 15.1　供应商信息查询窗体

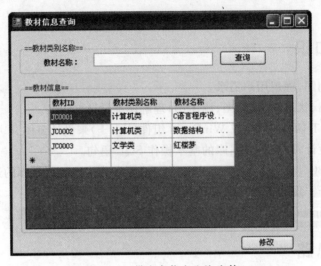

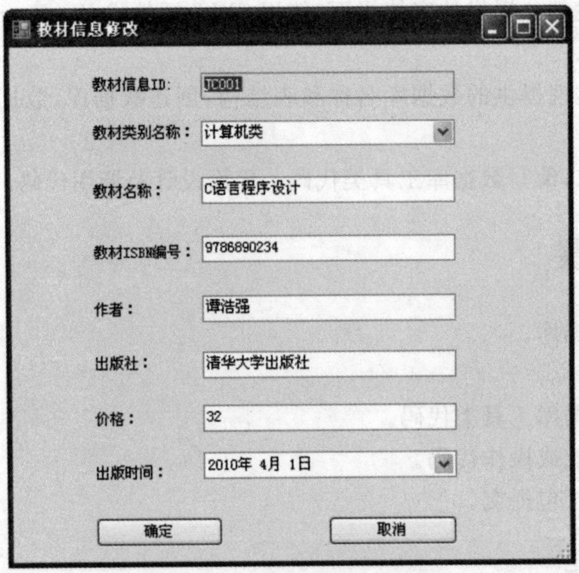

图 15.2　教材信息修改窗体

(2)在图 15.2 中修改教材相关信息，单击"确定"按钮，完成保存操作；单击"取消"按钮，取消修改。

3. 要求

(1)界面实现：实现图15.1、图15.2所示窗体。
(2)数据库实现：
1)创建数据库 HNIUEAMDB。
2)创建教材类别表(T_material_category)，其结构见表15.1。

表 15.1　教材类别表结构

字段名	字段说明	字段类型	是否允许为空	备注
Material_id	教材类别编号	Varchar(10)	否	主键
Material_name	教材类别名称	Varchar(50)	否	
Material_memo	备注	Text	是	

3)在表 T_material_category 中插入记录，见表15.2。

表 15.2　T_material_category 表记录

Material_id	Material_name	Material_name
JSJ001	计算机类	计算机
JXL002	机械类	机械
KPL004	科普类	科普
WXL003	文学类	文学

4)创建教材信息表(T_material_information)，其结构见表15.3。

表 15.3　教材信息表结构

字段名	字段说明	字段类型	是否允许为空	备注
Material_information_id	教材信息编号	Varchar(10)	否	主键
Material_category_id	教材类别编号	Varchar(10)	否	
Material__name	教材名称	Varchar(50)	否	
Material__ISBN	教材ISBN编号	Varchar(20)	否	
Author	作者	Varchar(20)	否	
Material_publisher	出版社	Varchar(50)	否	
Material_price	价格	Float	否	
Material_publication_date	出版时间	Datetime	否	

5)在表 T_material_information 中插入记录，见表15.4。

表 15.4　T_material_information 表记录

Material_ information_id	Material_ category_id	Material_ name	Material_ ISBN	Author	Material_ publisher	Material_ price	Material_ publication_ date
JC0001	JSJ001	C语言程序设计	9786890234	谭浩强	清华大学出版社	32	2010-04-1
JC0002	JSJ001	数据结构	9786589078	唐森宝	电子工业出版社	28	2008-07-1
JC0003	WXL003	红楼梦	9786510983	曹雪芹	机械工业出版社	90	2004-01-1

(3)功能实现：

1)功能需求如图 15.3 所示。

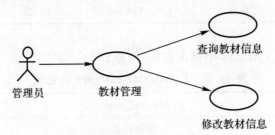

图 15.3　用例图

2)依据活动图完成教材信息修改功能,如图 15.4 所示。
3)依据活动图完成查询功能,如图 15.5 所示。
教材名称采用模糊查询,不输入教材名称时,显示所有教材信息。

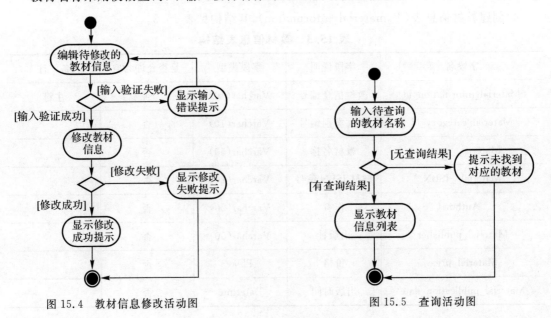

图 15.4　教材信息修改活动图　　　　图 15.5　查询活动图

项目十六　教务管理信息系统(三)

一、必备知识

(1)能读懂用例图,理解用户需求。
(2)能读懂类图、状态图、活动图、顺序图,理解详细设计。
(3)能使用 SWT 插件(java)或 WinForm 组件(C++)来设计窗体。
(4)数据库的设计,以及对数据库的基本操作 SQL 语句。
(5)能使用 JDBC(java)或 ADO.NET(C++)等方式建立与数据库的连接。
(6)能使用集合实现数据的存取和读出。
(7)能使用 Eclipse(java)或 Microsoft Visual Studio(C++)等开发工具并进行调试。

二、解题思路

(1)理解用例图、活动图。
(2)理解功能描述部分提供的窗体界面,使用 SWT 插件或 WinForm 设计窗体,以及窗体的布局和界面的控件。
(3)根据数据库实现提供的数据库名称和表结构,创建数据库、数据表、约束;并且在表中插入测试数据。
(4)根据功能要求,编写数据库工具类代码、界面设计及调用代码。

三、操作步骤

步骤一　创建数据库。
步骤二　界面设计。
步骤三　编写数据库工具类代码。
步骤四　编写功能或操作代码。
步骤五　按要求打包提交。

四、具体任务

1. 任务

你作为《教务管理信息系统》项目开发组的程序员,请实现下述功能:

- 查询教师信息；
- 删除教师信息。

2. 功能描述

(1)在图 16.1 中,输入教师名称,单击"查询"按钮时,将在教师列表中显示查询结果。选中一条教师信息后,单击"删除"按钮,完成删除操作。

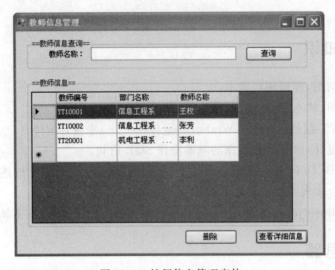

图 16.1　教师信息管理窗体

(2)在图 16.1 中选中一条教师信息后,单击"查看详细信息"按钮,打开教师信息窗体,并显示教师的详细信息,如图 16.2 所示。

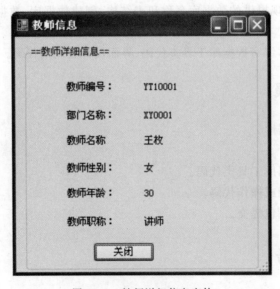

图 16.2　教师详细信息窗体

3. 要求

(1)界面实现。实现图 16.1、图 16.2 所示界面。

(2)数据库实现：

1)创建数据库 HNIUEAMDB。

2)创建部门表(T_department),其结构见表 16.1。

表 16.1 部门表结构

字段名	字段说明	字段类型	是否允许为空	备注
Department_id	教材部门编号	Varchar(10)	否	主键
Department_name	部门名称	Varchar(50)	否	
Department_memo	备注	Text	是	

3)在 T_department 中插入记录,见表 16.2。

表 16.2 T_department 表记录

Department_id	Department_name	Department_memo
XY0001	信息工程系	信息
XY0002	机电工程系	机电
XY0003	计算机工程系	计算机
XY0004	经济管理系	经济

3)创建教师信息表(T_teacher_information),其结构见表 16.3。

表 16.3 教师信息表(T_teacher_information)表结构

字段名	字段说明	字段类型	是否允许为空	备注
Teacher_id	教师信息编号	Varchar(10)	否	主键
Department_id	部门编号	Varchar(10)	否	
Teacher_name	教师名称	Varchar(8)	否	
Sex	性别	Bit	否	
Age	年龄	Int	否	
Prade	职称	Varchar(10)	否	

4)在 T_teacher_informationt 中插入记录,见表 16.4。

表 16.4 T_teacher_informationt 表记录

Teacher_id	Department_id	Teacher_name	Sex	Age	Prade
YT10001	XY0001	王枚	False	30	讲师
YT10002	XY0001	张芳	False	28	讲师
YT20001	XY0002	李利	True	45	教授

(3)功能实现:

1)功能需求如图 16.3 所示。

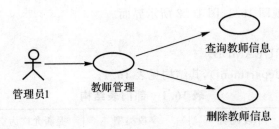

图 16.3 用例图

2)依据活动图完成教师信息删除功能,如图 16.4 所示。
3)依据活动图完成查询功能,如图 16.5 所示。

教师信息采用模糊查询,不输入教师名称时,显示所有教师信息。

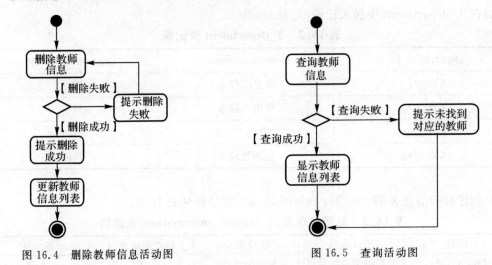

图 16.4 删除教师信息活动图　　　　图 16.5 查询活动图

项目十七　宿舍管理系统(一)

一、必备知识

(1)能读懂用例图,理解用户需求。
(2)能读懂类图、状态图、活动图、顺序图,理解详细设计。
(3)能使用 SWT 插件(java)或 WinForm 组件(C++)来设计窗体。
(4)数据库的设计,以及对数据库的基本操作 SQL 语句。
(5)能使用 JDBC(java)或 ADO.NET(C++)等方式建立与数据库的连接。
(6)能使用集合实现数据的存取和读出。
(7)能使用 Eclipse(java)或 Microsoft Visual Studio(C++)等开发工具并进行调试。

二、解题思路

(1)理解用例图、活动图。
(2)理解功能描述部分提供的窗体界面,使用 SWT 插件或 WinForm 设计窗体,以及窗体的布局和界面的控件。
(3)根据数据库实现提供的数据库名称和表结构,创建数据库、数据表、约束;并且在表中插入测试数据。
(4)根据功能要求,编写数据库工具类代码、界面设计及调用代码。

三、操作步骤

步骤一　创建数据库。
步骤二　界面设计。
步骤三　编写数据库工具类代码。
步骤四　编写功能或操作代码。
步骤五　按要求打包提交。

四、具体任务

1.任务

你作为《宿舍管理系统》项目组的程序员,请实现下述功能:

- 用户登录;
- 查询未住满的寝室。

2. 功能描述

(1)在图 17.1 中输入用户名和密码,单击"确定"按钮,打开宿舍管理系统主窗体,如图 17.2 所示。

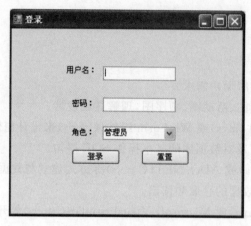

图 17.1 登录窗体

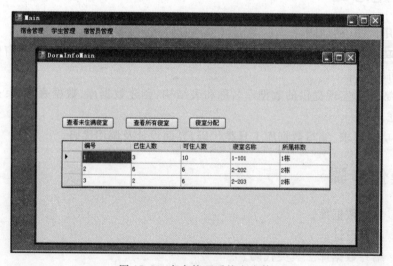

图 17.2 宿舍管理系统主窗体

(2)点击"宿舍管理"菜单,弹出 DormInfoMain 窗口(见图 17.2),单击"查看未住满寝室"按钮,显示查询结果(注:查询结果仍然显示在当前窗体),如图 17.3 所示。

3. 要求

(1)界面实现。实现图 17.1~图 17.3 所示界面。

项目十七 宿舍管理系统(一)

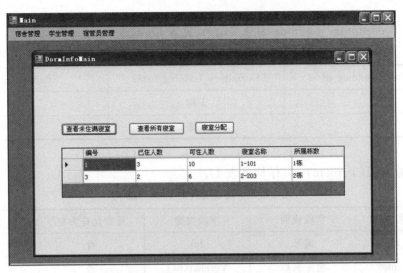

图 17.3 宿舍管理查询主窗体

(2)数据库实现:

1)创建数据库 DormDB。

2)创建管理员表(T_user),其结构如表 17.1。

表 17.1 管理员(T_user)表结构

字段名	字段说明	字段类型	是否允许为空	备 注
User_id	编号	Int	否	主键
User_name	登录名	Varchar(20)	否	
User_password	密码	Varchar(12)	否	
User_role	角色	Nchar(1)	否	1—管理员;2—宿管员;3—学生

3)在表 T_user 中插入记录,见表 17.2。

表 17.2 T_user 表记录

User_id	User_name	User_password	User_role
1	admin	admin	1

4)创建宿舍楼表(T_dormitory_building),其结构见表 17.3。

表 17.3 宿舍楼(T_dormitory_building)表结构

字段名	字段说明	字段类型	是否允许为空	备 注
Dormitory_building_id	编号	Int	否	主键
Dormitory_building_name	宿舍楼的名称	Varchar(64)	否	
Bed_Number	本栋楼的寝室的床位数	Int	否	本系统中假定位于同一栋楼里的寝室的床位数相同

5)在表 T_dormitory_building 中插入记录,见表 17.4。

表 17.4　T_dormitory_building 表记录

Dormitory_building_id	Dormitory_building_name	Bed_Number
1	1栋	10
2	2栋	6

6)创建寝室表(T_room),其结构见表 17.5。

表 17.5　寝室表(T_room)结构

字段名	字段说明	字段类型	是否允许为空	备注
Room_id	编号	Int	否	主键
Room_name	寝室名称	Varchar(64)	否	
Living_number	已住人数	Int	否	
Dormitory_building_id	所属宿舍楼编号	Int	否	外键

7)在表 T_room 中插入记录,见表 17.6。

表 17.6　T_room 表记录

Room_id	Room_name	Living_number	Dormitory_building_id
1	1-101	3	1
2	2-202	6	2
3	2-203	2	2

(3)功能实现:

1)功能需求如图 17.4 所示。

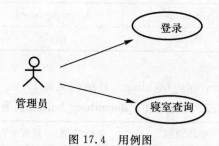

图 17.4　用例图

2)依据活动图完成管理员登录功能,如图 17.5 所示。

3)依据活动图完成"查询未住满寝室"功能,如图 17.6 所示。查询的条件是"已住人数 < 可住人数"。

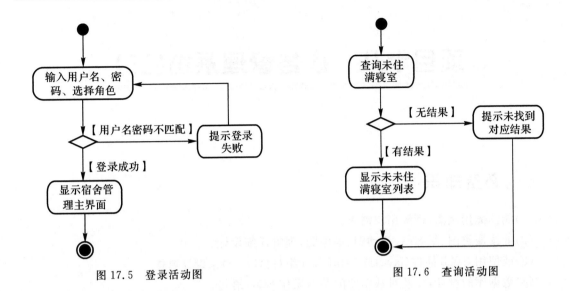

图 17.5　登录活动图　　　　　图 17.6　查询活动图

项目十八 宿舍管理系统(二)

一、必备知识

(1)能读懂用例图,理解用户需求。
(2)能读懂类图、状态图、活动图、顺序图,理解详细设计。
(3)能使用 SWT 插件(java)或 WinForm 组件(C++)来设计窗体。
(4)数据库的设计,以及对数据库的基本操作 SQL 语句。
(5)能使用 JDBC(java)或 ADO.NET(C++)等方式建立与数据库的连接。
(6)能使用集合实现数据的存取和读出。
(7)能使用 Eclipse(java)或 Microsoft Visual Studio(C++)等开发工具并进行调试。

二、解题思路

(1)理解用例图、活动图。
(2)理解功能描述部分提供的窗体界面,使用 SWT 插件或 WinForm 设计窗体,以及窗体的布局和界面的控件。
(3)根据数据库实现提供的数据库名称和表结构,创建数据库、数据表、约束;并且在表中插入测试数据。
(4)根据功能要求,编写数据库工具类代码、界面设计及调用代码。

三、操作步骤

步骤一　创建数据库。
步骤二　界面设计。
步骤三　编写数据库工具类代码。
步骤四　编写功能或操作代码。
步骤五　按要求打包提交。

四、具体任务

1. 任务

你作为《宿舍管理系统》项目开发组的程序员,请实现下述功能:

- 查询未分配寝室的学生。

2. 功能描述

(1)在图 18.1 中,选择"学生管理"菜单,将显示所有已分配寝室的学生信息,如图 18.2 所示。

图 18.1　主窗体

(2)在图 18.2 中单击"查看未分配学生"按钮,将显示如图 18.3 所示的查询结果(注意查询结果仍然显示在当前界面里)。

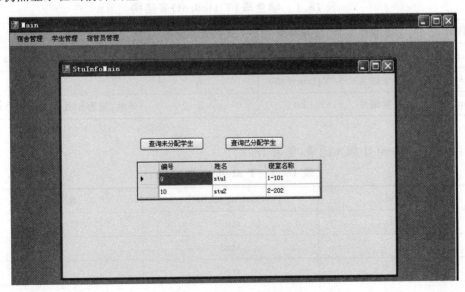

图 18.2　学生管理窗体

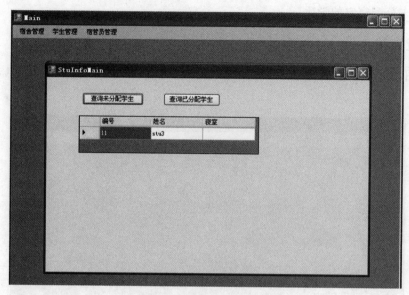

图 18.3　查询结果窗体

3. 要求

(1)界面实现。实现图 18.1～图 18.3 所示界面。

(2)数据库实现：

1)创建数据库 DormDB。

2)创建学生表(T_student)，其结构见表 18.1。

表 18.1　学生表(T_student)表结构

字段名	字段说明	字段类型	是否允许为空	备注
Student_id	编号	Int	否	主键
Student_name	学生姓名	Varchar(20)	否	
Room_id	寝室编号	Int	是	外键；值为 null 时表示未分配寝室

3)在表 T_student 中插入记录，见表 18.2。

表 18.2　T_student 表记录

Student_id	Student_name	Room_id
1	stu1	1
2	stu2	2
3	stu3	null

4)创建宿舍楼表(T_Dormitory_building)，其结构见表 18.3。

表 18.3 宿舍楼表(T_Dormitory_building)表结构

字段名	字段说明	字段类型	是否允许为空	备注
Dormitory_building_id	编号	Int	否	主键
Dormitory_building_name	宿舍楼的名称	Varchar(64)	否	
Bed_Number	本栋楼的寝室的床位数	Int	否	本系统中假定位于同一栋楼里的寝室的床位数相同

5)在表 T_Dormitory_building 中插入记录,见表 18.4。

表 18.4 T_Dormitory_building 表记录

Dormitory_building_id	Dormitory_building_name	Bed_Number
1	1栋	10
2	2栋	6

6)创建寝室表(T_room),表结构见表 18.5。

表 18.5 寝室表结构

字段名	字段说明	字段类型	是否允许为空	备注
Room_id	寝室编号	Int	否	主键
Room_name	寝室名称	Varchar(64)	否	
Living_number	已住人数	Int	否	
Dormitory_building_id	所属宿舍楼编号	Int	否	外键

7)在表 T_room 中插入记录,见表 18.6。

表 18.6 T_room 表记录

Room_id	Room_name	Living_number	Dormitory_building_id
1	1-101	3	1
2	2-202	6	2
3	2-203	2	2

(3)功能实现:

1)功能需求如图 18.4 所示。

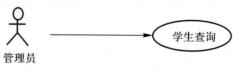

图 18.4 用例图

2) 依据活动图完成"已分配寝室学生"查询功能,如图 18.5 所示。

3) 依据活动图完成"查询未分配学生"功能,如图 18.6 所示。

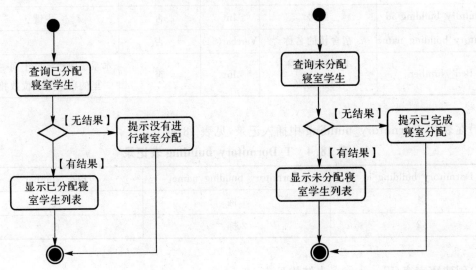

图 18.5 "已分配寝室学生"查询活动图　　图 18.6 "未分配学生"查询活动图

项目十九　宿舍管理系统(三)

一、必备知识

(1)能读懂用例图,理解用户需求。
(2)能读懂类图、状态图、活动图、顺序图,理解详细设计。
(3)能使用 SWT 插件(java)或 WinForm 组件(C++)来设计窗体。
(4)数据库的设计,以及对数据库的基本操作 SQL 语句。
(5)能使用 JDBC(java)或 ADO.NET(C++)等方式建立与数据库的连接。
(6)能使用集合实现数据的存取和读出。
(7)能使用 Eclipse(java)或 Microsoft Visual Studio(C++)等开发工具并进行调试。

二、解题思路

(1)理解用例图、活动图。
(2)理解功能描述部分提供的窗体界面,使用 SWT 插件或 WinForm 设计窗体,以及窗体的布局和界面的控件。
(3)根据数据库实现提供的数据库名称和表结构,创建数据库、数据表、约束;并且在表中插入测试数据。
(4)根据功能要求,编写数据库工具类代码、界面设计及调用代码。

三、操作步骤

步骤一　创建数据库。
步骤二　界面设计。
步骤三　编写数据库工具类代码。
步骤四　编写功能或操作代码。
步骤五　按要求打包提交。

四、具体任务

1.任务

你作为《宿舍管理系统》项目开发组的程序员,请实现下述功能:

- 用户登录；
- 分配宿管员。

2. 功能描述

（1）在图 19.1 中输入用户名和密码，单击"确定"按钮，打开宿舍管理系统主窗体，如图 19.2 所示。

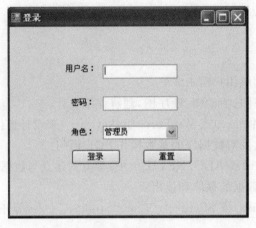

图 19.1　登录窗体

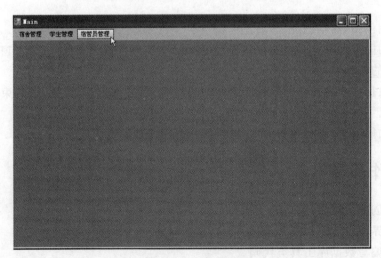

图 19.2　主窗体

（2）在图 19.2 中，单击"宿管员管理"菜单项。

（3）在图 19.3 中，选择未分配宿管员的宿舍楼，然后选择宿管员。

（4）单击图 19.3 中的"分配"按钮，完成宿管员分配操作（注：宿管员分配，就是修改 T_room_manager 表的 Dormitory_building_id 的值）。

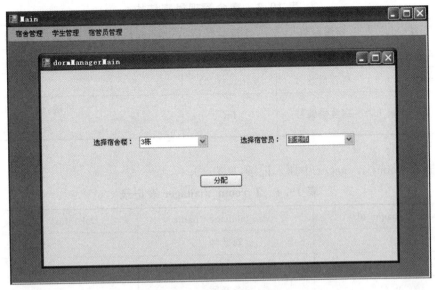

图 19.3 宿管员管理窗体

3. 要求

(1)界面实现。实现图 19.1～图 19.3 所示窗体。

(2)数据库实现：

1)创建数据库 DormDB。

2)创建管理员表(T_user)，其结构如表 19.1。

表 19.1 管理员(T_user)表结构

字段名	字段说明	字段类型	是否允许为空	备注
User_id	编号	Int	否	主键
User_name	登录名	Varchar(20)	否	
User_password	密码	Varchar(12)	否	
User_role	角色	Nchar(1)	否	1—管理员；2—宿管员；3—学生

3)在表 T_user 中插入记录，见表 19.2。

表 19.2 T_user 表记录

User_id	User_name	User_password	User_role
1	admin	admin	1

4)创建宿舍管理员表(T_room_manager)，其结构见表 19.3。

表 19.3　宿舍管理员表结构

字段名	字段说明	字段类型	是否允许为空	备注
room_manager_id	编号	Int	否	主键
room_manager_name	宿管员姓名	Varchar(20)	否	
Dormitory_building_id	宿舍楼编号	Int	是	外键,值为 null 时表示未分配宿舍楼

5)在表 T_room_manager 中插入记录,见表 19.4。

表 19.4　T_room_manager 表记录

room_manager_id	room_manager_name	Dormitory_building_id
1	刘老师	1
2	王老师	2
3	李老师	null

6)创建宿舍楼表(D_dormitory_building),其结构见表 19.5。

表 19.5　D_dormitory_building 表记录

字段名	字段说明	字段类型	是否允许为空	备注
Dormitory_building_id	编号	Int	否	主键
Dormitory_building_name	宿舍楼的名称	Varchar(64)	否	
Bed_Number	本栋楼的寝室的床位数	Int	否	本系统中假定位于同一栋楼里的寝室的床位数相同

7)在表 D_dormitory_building 中插入记录,见表 19.6。

表 19.6　D_dormitory_building 表记录

Dormitory_building_id	Dormitory_building_name	Bed_Number
1	1栋	10
2	2栋	6
3	3栋	8

(3)功能实现:

1)功能需求如图 19.4 所示。

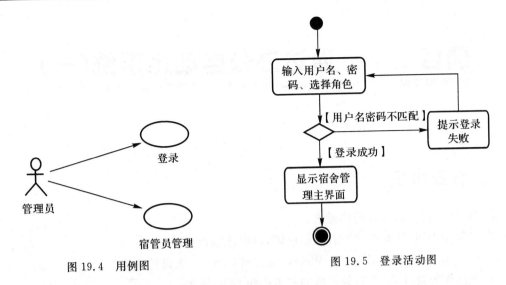

图 19.4 用例图　　　　图 19.5 登录活动图

2)依据活动图完成管理员登录功能,如图 19.5 所示。

3)依据活动图完成"宿管员管理"功能,如图 19.6 所示。

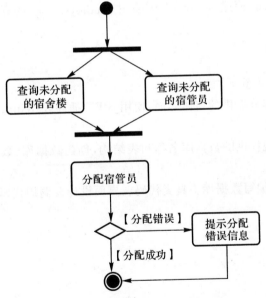

图 19.6 "宿管员管理"查询活动图

项目二十 通达办公自动化系统(一)

一、必备知识

(1)能读懂用例图,理解用户需求。
(2)能读懂类图、状态图、活动图、顺序图,理解详细设计。
(3)能使用SWT插件(java)或WinForm组件(C++)来设计窗体。
(4)数据库的设计,以及对数据库的基本操作SQL语句。
(5)能使用JDBC(java)或ADO.NET(C++)等方式建立与数据库的连接。
(6)能使用集合实现数据的存取和读出。
(7)能使用Eclipse(java)或Microsoft Visual Studio(C++)等开发工具并进行调试。

二、解题思路

(1)理解用例图、活动图。
(2)理解功能描述部分提供的窗体界面,使用SWT插件或WinForm设计窗体,以及窗体的布局和界面的控件。
(3)根据数据库实现提供的数据库名称和表结构,创建数据库、数据表、约束;并且在表中插入测试数据。
(4)根据功能要求,编写数据库工具类代码、界面设计及调用代码。

三、操作步骤

步骤一　创建数据库。
步骤二　界面设计。
步骤三　编写数据库工具类代码。
步骤四　编写功能或操作代码。
步骤五　按要求打包提交。

四、具体任务

1.任务

你作为《通达办公自动化系统》项目开发组的程序员,请实现下述功能:

- 添加办公用品类别；
- 查询办公用品。

2. 功能描述

(1) 在图 20.1 中，单击"类别添加"按钮，打开"类别添加"窗体，如图 20.2 所示。

图 20.1 主窗体

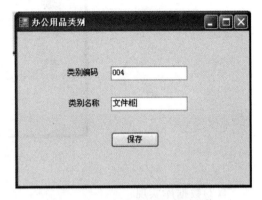

图 20.2 添加办公用品类别窗体

(2) 点击"类别添加"按钮，进入"办公用品类别窗口"(见图 20.2)，输入类别编码和类别名称，单击"保存"按钮，完成类别添加功能，并返回主窗体。

(3) 在图 20.3 中，输入办公用品编号或办公用品名称，单击"查询"按钮后，则在办公用品信息窗体中显示查询结果。如图 20.3 所示。

图 20.3 办公用品查询窗体

(4) 在图 20.3 中任选一条办公用品记录，打开办公用品详细信息窗体，并显示办公用品详细信息，如图 20.4 所示。

3. 要求

(1) 界面实现。实现图 20.1～图 20.4 所示窗体。

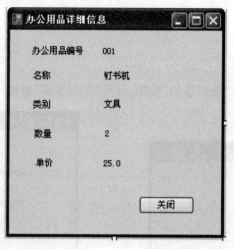

图 20.4 办公用品详细信息窗体

(2)数据库实现:
1)创建数据库 OADB。
2)创建办公用品类别表(T_category),其结构见表 20.1。

表 20.1 办公用品类别表结构

字段名	字段说明	字段类型	是否允许为空	备 注
Category_Id	办公用品类别编号	Varchar(20)	否	主键
Category_Name	办公用品类别名称	Varchar(20)	否	

3)在人表 T_category 中插入记录,见表 20.2。

表 20.2 T_category 表记录

Category_Id	Category_Name
001	文具
002	耗材
003	纸张

4)创建办公用品表(T_product),其结构见表 20.3。

表 20.3 办公用品表结构

字段名	字段说明	字段类型	是否允许为空	备 注
Product_id	办公用品编号	Varchar(20)	否	主键
Product_name	名称	Varchar(20)	否	
Category_Id	类别	Varchar(20)	否	外键
Product_number	数量	Float	否	
Product_price	单价	Float	否	

5)在入表 T_product 中插入记录,见表 20.4。

表 20.4 T_product 表记录

Product_id	Product_name	Category_id	Product_number	Product_price
001	签字笔	001	2	1.0
002	2B 铅笔	001	4	1.5
003	钉书机	001	5	25.0

(3)功能实现:

1)功能需求如图 20.5 所示。

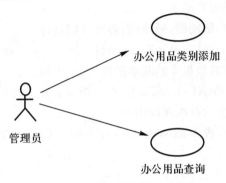

图 20.5 办公用品管理用例图

2)依据活动图完成办公用品类别添加功能,如图 20.6 所示。

3)依据活动图完成查询功能,如图 20.7 所示。办公用品编号和名称采用模糊查询,不输入办公用品编号和名称时,显示所有办公用品。

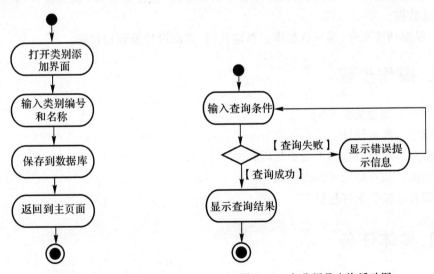

图 20.6 办公用品类别添加活动图　　图 20.7 办公用品查询活动图

项目二十一　通达办公自动化系统(二)

一、必备知识

(1)能读懂用例图,理解用户需求。
(2)能读懂类图、状态图、活动图、顺序图,理解详细设计。
(3)能使用 SWT 插件(java)或 WinForm 组件(C++)来设计窗体。
(4)数据库的设计,以及对数据库的基本操作 SQL 语句。
(5)能使用 JDBC(java)或 ADO.NET(C++)等方式建立与数据库的连接。
(6)能使用集合实现数据的存取和读出。
(7)能使用 Eclipse(java)或 Microsoft Visual Studio(C++)等开发工具并进行调试。

二、解题思路

(1)理解用例图、活动图。
(2)理解功能描述部分提供的窗体界面,使用 SWT 插件或 WinForm 设计窗体,以及窗体的布局和界面的控件。
(3)根据数据库实现提供的数据库名称和表结构,创建数据库、数据表、约束;并且在表中插入测试数据。
(4)根据功能要求,编写数据库工具类代码、界面设计及调用代码。

三、操作步骤

步骤一　创建数据库。
步骤二　界面设计。
步骤三　编写数据库工具类代码。
步骤四　编写功能或操作代码。
步骤五　按要求打包提交。

四、具体任务

1.任务

你作为《通达办公自动化系统》项目开发组的程序员,请实现下述功能:

- 添加会议室；
- 查询会议室预订。

2. 功能描述

(1) 在图 21.1 中，单击主界面上的"会议室添加"按钮，打开"会议室"窗口，如图 21.2 所示。

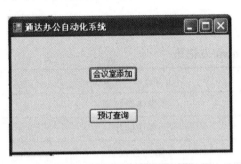

图 21.1　会议室管理主界面

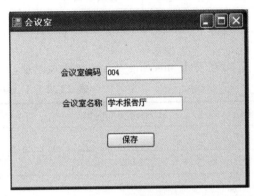

图 21.2　会议室添加界面

(2) 在图 21.2 中，输入会议室编码和会议室名称，单击"保存"按钮保存会议室信息，并返回主窗体。

(3) 单击住界面的"预定查询"按钮，进入"会议室预定查询"窗口（图 21.3），输入会议室或预订人，单击"查询"按钮，将会显示查询结果。

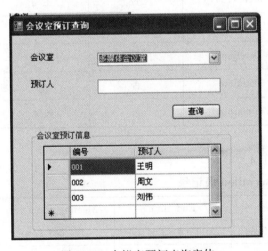

图 21.3　会议室预订查询窗体

图 21.4　会议室预订详细信息窗体

(4) 在图 21.3 中选择某条会议室预订信息，在图 21.4 中显示开会议室预订详细信息界面。

3. 要求

(1) 界面实现。实现图 21.1～图 21.4 所示界面。

(2)数据库实现:

1)创建数据库 OADB。

2)创建会议室表(T_meeting_room),其结构见表 21.1。

表 21.1　会议室表结构

字段名	字段说明	字段类型	是否允许为空	备注
Meeting_room_id	会议室编号	Varchar(20)	否	主键
Meeting_room_name	会议室名称	Varchar(20)	否	

3)在表 T_meeting_room 中插入记录,见表 21.2。

表 21.2　T_meeting_room 表记录

Meeting_room_id	Meeting_room_name
001	多媒体会议室
002	多功能厅
003	第三会议室

4)创建会议室预订表(T_reservation),其结构见表 21.3。

表 21.3　会议室预订表结构

字段名	字段说明	字段类型	是否允许为空	备注
Reservation_id	会议室预订编号	Varchar(20)	否	主键
Reservation_name	预订人	Varchar(20)	否	
Start_time	开始时间	Datetime	否	
End_time	结束时间	Datetime	否	
Meeting_room_id	会议室编号	Varchar(20)	否	外键

5)在表 T_reservation 中插入记录,见表 21.4。

表 21.4　T_reservation 表记录

Reservation_id	Reservation_name	Start_time	End_time	Meeting_room_id
001	王明	2011-6-3 15:30	2011-6-3 17:50	001
002	周文	2011-6-4 8:30	2011-6-4 9:50	001
003	刘伟	2011-6-5 10:00	2011-6-5 12:00	001

(3)功能实现:

1)功能需求如图 21.5 所示。

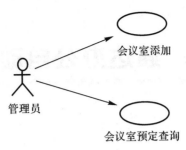

图 21.5　会议室管理用例图

2）依据活动图完成会议室添加功能,如图 21.6 所示。

3）依据活动图完成查询功能,如图 21.7 所示。会议室通过下拉列表框选择输入,不输入预订人时,显示所有该会议室的预订记录。输入预订人时,可以模糊查询该预订人的所有预订记录。

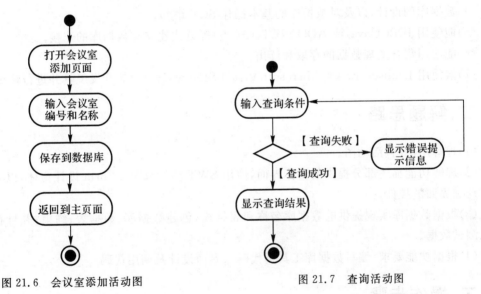

图 21.6　会议室添加活动图　　　　　　图 21.7　查询活动图

项目二十二　通达办公自动化系统(三)

一、必备知识

(1)能读懂用例图,理解用户需求。
(2)能读懂类图、状态图、活动图、顺序图,理解详细设计。
(3)能使用SWT插件(java)或WinForm组件(C++)来设计窗体。
(4)数据库的设计,以及对数据库的基本操作SQL语句。
(5)能使用JDBC(java)或ADO.NET(C++)等方式建立与数据库的连接。
(6)能使用集合实现数据的存取和读出。
(7)能使用Eclipse(java)或Microsoft Visual Studio(C++)等开发工具并进行调试。

二、解题思路

(1)理解用例图、活动图。
(2)理解功能描述部分提供的窗体界面,使用SWT插件或WinForm设计窗体,以及窗体的布局和界面的控件。
(3)根据数据库实现提供的数据库名称和表结构,创建数据库、数据表、约束;并且在表中插入测试数据。
(4)根据功能要求,编写数据库工具类代码、界面设计及调用代码。

三、操作步骤

步骤一　创建数据库。
步骤二　界面设计。
步骤三　编写数据库工具类代码。
步骤四　编写功能或操作代码。
步骤五　按要求打包提交。

四、具体任务

1. 任务

你作为《通达办公自动化系统》项目组的程序员,请实现下述功能:

- 添加部门；
- 查询员工信息。

2. 功能描述

(1)在图 22.1 中,单击主界面上的"部门添加"按钮,打开"部门"窗体。

图 22.1　主窗体

图 22.2　部门添加窗体

(2)在图 22.2 中,输入部门编码和部门名称,单击"保存"按钮,完成部门添加操作,并返回主窗体。

(3)点击主界面"员工信息查询"按钮,进入"员工信息查询"窗口(见图 22.3),输入部门或员工姓名,单击"查询"按钮后。则在员工信息列表中显示查询结果。

图 22.3　员工信息查询窗体

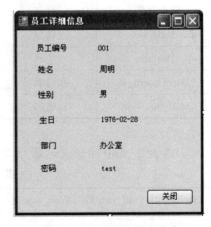

图 22.4　员工详细信息窗体

(4)在图 22.3 中,任选一条员工信息记录,打开员工详细信息窗体,并显示该员工的详细信息,如图 22.4 所示。

3. 要求

(1)界面实现。实现图 22.1～图 22.4 所示界面。

(2)数据库实现:

1)创建数据库 OADB。

2)创建部门表(T_department),其结构见表 22.1。

表 22.1 部门表结构

字段名	字段说明	字段类型	是否允许为空	备注
Department_id	部门编号	Varchar(20)	否	主键
Department_name	部门名称	Varchar(20)	否	

3)在表 T_department 中插入记录,见表 22.2。

表 22.2 T_department 表记录

Department_id	Department_name
001	办公室
002	财务处
003	人事处

4)创建管理员表(T_staff),其结构见表 22.3。

表 22.3 管理员表结构

字段名	字段说明	字段类型	是否允许为空	备注
Staff_id	员工编号	Varchar(20)	否	主键
Staff_name	姓名	Varchar(20)	否	
Staff_sex	性别	Varchar(2)	否	
Birthday	生日	Datetime	否	
Department_id	部门编号	Varchar(20)	否	外键
Staff_password	密码	Varchar(20)	否	

5)在表 T_staff 中插入记录,见表 22.4。

表 22.4 T_staff 表记录

Staff_id	Staff_name	Staff_sex	Birthday	Department_id	Staff_password
001	周明	男	1976-02-28	001	Test
002	李青	女	1981-03-04	001	Test
003	刘欣	男	1992-12-25	001	test

(3)功能实现:

1)功能需求如图 22.5 所示。

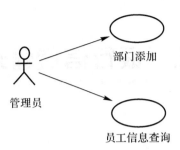

图 22.5 员工管理用例图

2)依据活动图完成部门添加功能,如图 22.6 所示。

3)依据活动图完成查询功能,如图 22.7 所示。部门通过下拉列表框选择输入,不输入员工姓名时,显示所有该部门的员工。输入员工姓名时,可以模糊查询相关的员工记录。

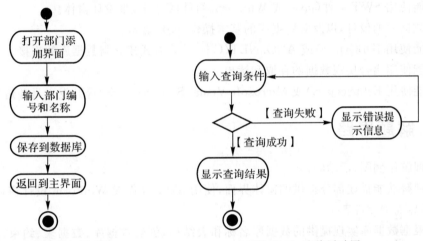

图 22.6 部门添加活动图   图 22.7 查询活动图

项目二十三　银行信贷管理系统(一)

一、必备知识

(1)能读懂用例图,理解用户需求。
(2)能读懂类图、状态图、活动图、顺序图,理解详细设计。
(3)能使用 SWT 插件(java)或 WinForm 组件(C++)来设计窗体。
(4)数据库的设计,以及对数据库的基本操作 SQL 语句。
(5)能使用 JDBC(java)或 ADO.NET(C++)等方式建立与数据库的连接。
(6)能使用集合实现数据的存取和读出。
(7)能使用 Eclipse(java)或 Microsoft Visual Studio(C++)等开发工具并进行调试。

二、解题思路

(1)理解用例图、活动图。
(2)理解功能描述部分提供的窗体界面,使用 SWT 插件或 WinForm 设计窗体,以及窗体的布局和界面的控件。
(3)根据数据库实现提供的数据库名称和表结构,创建数据库、数据表、约束;并且在表中插入测试数据。
(4)根据功能要求,编写数据库工具类代码、界面设计及调用代码。

三、操作步骤

步骤一　创建数据库。
步骤二　界面设计。
步骤三　编写数据库工具类代码。
步骤四　编写功能或操作代码。
步骤五　按要求打包提交。

四、具体任务

1. 任务

你作为《银行信贷管理系统》项目组的程序员,请实现下述功能:

- 添加客户信息；
- 查询所有客户信息。

2. 功能描述

(1)查询所有客户信息。在图 23.1 中，点击"查询"按钮，显示所有客户信息，如图 23.2 所示。

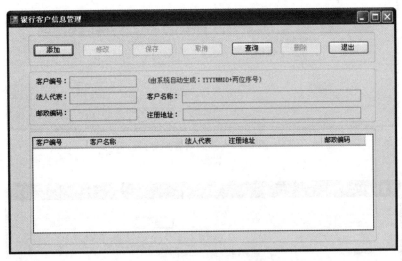

图 23.1　银行客户信息管理界面

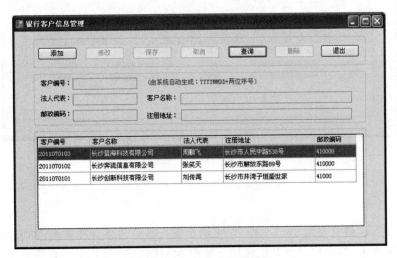

图 23.2　银行客户查询界面

(2)添加客户信息。在图 23.1 中，点击"添加"按钮显示图 23.3 所示界面。

在图 23.3 中输入所有客户信息，点击"保存"按钮，客户列表刷新，"保存"、"取消"按钮变成不可用状态，如图 23.4 所示。

桌面应用开发基础

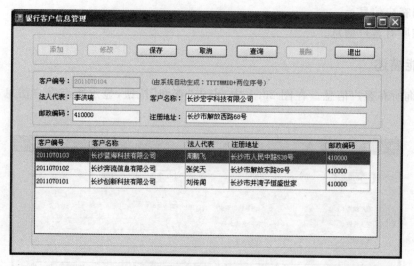

图 23.3　添加新客户界面

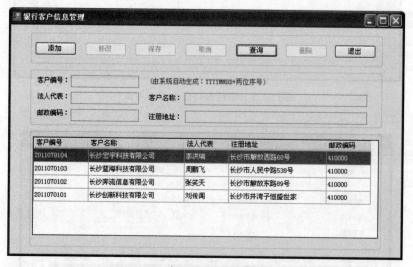

图 23.4　添加新客户成功后界面

3. 要求

(1)界面实现。请实现图 23.1～图 23.4 所示银行客户信息管理界面,要求如下:

1)在图 23.3 中,客户编号由系统按照 YYYYMMDD＋两位序号的规则自动生成。

2)"查询"、"退出"按钮始终处于可用状态,"客户编号"文本框始终处于不可用状态,"添加"按钮初始化为可用状态,其他控件初始化为非可用状态。

3)单击"添加"按钮,"保存""取消"按钮,输入文本框可用,"添加""修改""删除"按钮不可用。

4)单击"保存"按钮,"添加"按钮可用,"保存""取消""修改""删除"按钮、输入文本框没内容清空并不可用。

5)单击"取消"按钮,"保存""取消""修改""删除"按钮、输入文本框没内容清空并不可用。
6)单击"退出"按钮,关闭窗体。
(2)数据库实现:
1)创建数据库 BankCreditLoanDB。
2)创建客户基本信息表 T_customer_info,其结构见表 23.1。

表 23.1 客户基本信息表(T_customer_info)表结构

字段名	字段说明	字段类型	是否允许为空	备注
Cust_id	客户编号	Char(10)	否	主键
Cust_name	客户名称	Varchar(60)	否	
Legal_name	法人代表	Varchar(10)	否	
Reg_address	注册地址	Varchar(60)	否	
Post_code	邮政编码	Char(6)	否	

3)在表 T_customer_info 中插入记录,见表 23.2。

表 23.2 客户基本信息表(T_customer_info)记录

Cust_id	Cust_name	Legal_name	Reg_address	Post_code
2011070101	××创新科技有限公司	刘××	××市井湾子恒盛世家	410000
2011070102	××奔流信息有限公司	张××	××市解放东路 89 号	410000
2011070103	××蓝海科技有限公司	周××	××市人民中路 538 号	410000

(3)功能实现:

1)功能需求如图 23.5 所示。

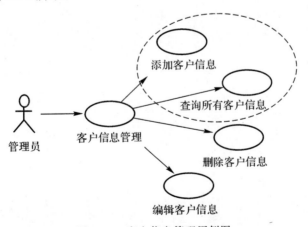

图 23.5 客户信息管理用例图

2)依据活动图完成查询所有客户信息功能,如图 23.6 所示。
3)依据活动图完成添加客户信息功能,如图 23.7 所示。

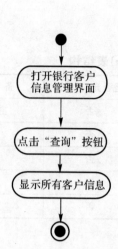
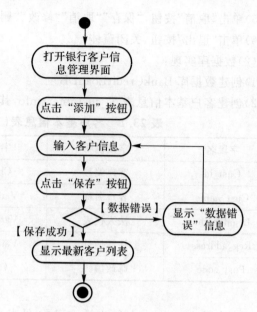

图 23.6 显示所有客户信息活动图　　图 23.7 添加客户信息活动图

项目二十四 银行信贷管理系统(二)

一、必备知识

(1)能读懂用例图,理解用户需求。
(2)能读懂类图、状态图、活动图、顺序图,理解详细设计。
(3)能使用SWT插件(java)或WinForm组件(C++)来设计窗体。
(4)数据库的设计,以及对数据库的基本操作SQL语句。
(5)能使用JDBC(java)或ADO.NET(C++)等方式建立与数据库的连接。
(6)能使用集合实现数据的存取和读出。
(7)能使用Eclipse(java)或Microsoft Visual Studio(C++)等开发工具并进行调试。

二、解题思路

(1)理解用例图、活动图。
(2)理解功能描述部分提供的窗体界面,使用SWT插件或WinForm设计窗体,以及窗体的布局和界面的控件。
(3)根据数据库实现提供的数据库名称和表结构,创建数据库、数据表、约束;并且在表中插入测试数据。
(4)根据功能要求,编写数据库工具类代码、界面设计及调用代码。

三、操作步骤

步骤一 创建数据库。
步骤二 界面设计。
步骤三 编写数据库工具类代码。
步骤四 编写功能或操作代码。
步骤五 按要求打包提交。

四、具体任务

1.任务

你作为承接《银行信贷管理系统》项目组的程序员,请实现下述功能:

- 查看贷款信息；
- 修改贷款信息。

2. 功能描述

（1）查看贷款信息。在图24.1中，点击左边客户列表中客户名称，所点客户的贷款信息显示在右边的贷款列表中，如图24.2所示。

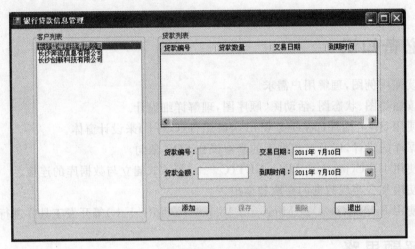

图24.1　银行贷款信息管理界面

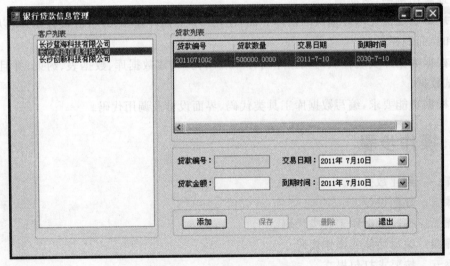

图24.2　查看客户贷款界面

（2）修改贷款信息。在图24.2中，点击贷款列表中的贷款编号显示图24.3所示界面。在图24.3中修改贷款信息，点击"保存"按钮，贷款列表刷新，如图24.4所示。

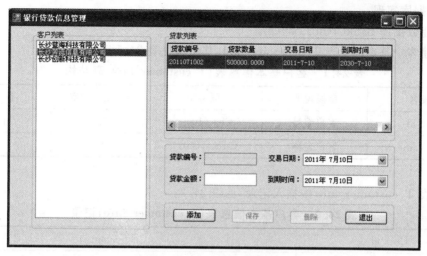

图 24.3　修改客户贷款界面

图 24.4　修改客户贷款成功后界面

3. 要求

（1）界面实现。请实现图 24.1～图 24.4 所示银行贷款信息管理界面。要求如下：

1）打开银行贷款信息管理界面时，客户名称动态加载到客户列表中，"保存""删除"按钮为不可用状态，如图 24.1 所示；

2）点击左边客户列表中客户名称，所点客户的贷款信息显示在右边的贷款列表中，如图 24.2 所示；

3）点击贷款列表中贷款编号，该编号的贷款信息显示在贷款列表下面的文本框控件中，"添加"按钮不可用，"保存""删除"按钮可用，如图 24.3 所示；

4）修改贷款信息后，点击"保存"按钮保存后，"添加"按钮可用，"保存""删除"按钮不可用，如图 24.4 所示。

(2)数据库实现:
1)创建数据库 BankCreditLoanDB。
2)创建客户基本信息表 T_customer_info,其结构见表 24.1。

表 24.1 客户基本信息表(T_customer_info)表结构

字段名	字段说明	字段类型	是否允许为空	备注
Cust_id	客户编号	Char(10)	否	主键
Cust_name	客户名称	Varchar(60)	否	

3)在表 T_customer_info 中插入记录,见表 24.2。

表 24.2 客户基本信息表(T_customer_info)记录

Cust_id	Cust_name
2011070101	长沙创新科技有限公司
2011070102	长沙奔流信息有限公司
2011070103	长沙蓝海科技有限公司

4)创建客户贷款明细表 T_loan_detail,其结构见表 24.3。

表 24.3 客户贷款明细表(T_loan_detail)表结构

字段名	字段说明	字段类型	是否允许为空	备注
Loan_id	贷款编号	Char(10)	否	主键
Cust_id	客户编号	Char(10)	否	外键
Trans_date	交易日期	Datetime	否	
Loan_amount	贷款金额	Money	否	
Expire_time	到期时间	Datetime	否	

5)在表 T_loan_detail 中插入记录,见表 24.4。

表 24.4 客户贷款明细表(T_loan_detail)记录

Loan_id	Cust_id	Trans_date	Loan_amount	Expire_time
2011071001	2011070101	2011-07-10	100000	2016-07-10
2011071002	2011070102	2011-07-10	200000	2020-07-10
2011071003	2011070103	2011-07-10	300000	2025-07-10

(3)功能实现:
1)功能需求如图 24.5 所示。

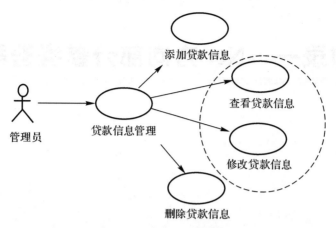

图 24.5 贷款信息管理用例图

2）依据活动图完成查看贷款信息功能，如图 24.6 所示。
3）依据活动图完成查看修改贷款信息功能，如图 24.7 所示。

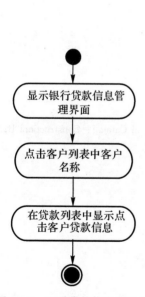

图 24.6 查看贷款信息活动图

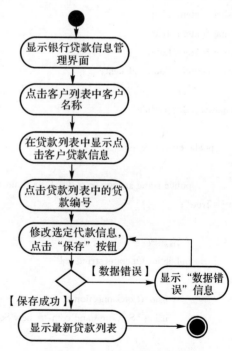

图 24.7 修改贷款信息活动图

附录一 Net 方向部分参考答案

项目一

```csharp
/**
*数据库操作公共类
*/
using System;
using System.Collections.Generic;
using System.Linq;
using System.Text;
using System.Data;
using System.Data.SqlClient;

namespace ProjectInfo
{
    public class DBHelper
    {
        public static string str = @"Data Source=localhost\SA;Initial Catalog=ConstructionDB;Integrated Security=True";

        //增删改
        public int Operate(string sql)
        {
            using(SqlConnection con = new SqlConnection(str))
            using (SqlCommand com = new SqlCommand(sql, con))
            {
                con.Open();
                return com.ExecuteNonQuery();
            }
        }

        //查询单个值
        public object GetScalar(string sql)
        {
            using (SqlConnection con = new SqlConnection(str))
```

```csharp
            using (SqlCommand com = new SqlCommand(sql, con))
            {
                con.Open();
                return com.ExecuteScalar();
            }
        }

        //查询结果集
        public SqlDataReader GetReader(string sql)
        {
            SqlConnection con = new SqlConnection(str);
            SqlCommand com = new SqlCommand(sql, con);
            {
                con.Open();
                return com.ExecuteReader(CommandBehavior.CloseConnection);
            }
        }

        public SqlDataAdapter da;
        public void FillData(string sql, DataSet ds)
        {
            SqlConnection con = new SqlConnection(str);
            da = new SqlDataAdapter(sql, con);
            da.Fill(ds);
        }
    }
}

/**
 *登录界面操作类
 **/
using System;
using System.Collections.Generic;
using System.ComponentModel;
using System.Data;
using System.Drawing;
using System.Linq;
using System.Text;
using System.Windows.Forms;

namespace ProjectInfo
{
    public partial class frmLogin : Form
```

```csharp
    {
        public frmLogin()
        {
            InitializeComponent();
        }

        private void button2_Click(object sender, EventArgs e)
        {
            Application.Exit();
        }

        //登录
        private void button1_Click(object sender, EventArgs e)
        {
            DBHelper db = new DBHelper();
            string user = txtUser.Text.Trim();
            string pwd = txtPwd.Text.Trim();

            string sql = string.Format("select M_password from T_manager where M_id = '{0}'", user);
            string realPwd = Convert.ToString(db.GetScalar(sql));
            if (pwd == realPwd)
            {
                frmSearchProject frm = new frmSearchProject();
                frm.Show();
                this.Hide();
            }
            else
            {
                MessageBox.Show("用户名或密码有误!", "系统提示", MessageBoxButtons.OK, MessageBoxIcon.Error);
                return;
            }

        }
    }
}

/**
 * 工程详细信息类
 * */
using System;
using System.Collections.Generic;
using System.ComponentModel;
```

```csharp
using System.Data;
using System.Drawing;
using System.Linq;
using System.Text;
using System.Windows.Forms;
using System.Data.SqlClient;

namespace ProjectInfo
{
    public partial class frmProjectInfo : Form
    {
        public string id = "";
        public frmProjectInfo()
        {
            InitializeComponent();
        }

        private void button1_Click(object sender, EventArgs e)
        {
            this.Close();
        }

        private void frmProjectInfo_Load(object sender, EventArgs e)
        {
            string sql = string.Format("select * from T_project where project_id = '{0}'", id);
            DBHelper db = new DBHelper();
            SqlDataReader dr = db.GetReader(sql);
            while (dr.Read())
            {
                lblId.Text = id;
                lblName.Text = Convert.ToString(dr["project_name"]);
                lblInvi_dept.Text = Convert.ToString(dr["Invi_dept"]);
                lblType.Text = Convert.ToString(dr["System_type"]);
                lblPhone.Text = Convert.ToString(dr["Telephone"]);
                lblState.Text = Convert.ToString(dr["System_state"]);

            }
            dr.Close();
        }
    }
}
```

```csharp
/**
 * 工程查询界面类
 **/
using System;
using System.Collections.Generic;
using System.ComponentModel;
using System.Data;
using System.Drawing;
using System.Linq;
using System.Text;
using System.Windows.Forms;

namespace ProjectInfo
{
    public partial class frmSearchProject : Form
    {
        DataSet ds = new DataSet();
        DBHelper db = new DBHelper();
        public frmSearchProject()
        {
            InitializeComponent();
        }

        private void frmSearchProject_Load(object sender, EventArgs e)
        {
            cmbState.SelectedIndex = 0;
        }

        private void button1_Click(object sender, EventArgs e)
        {
            if (ds.Tables.Count != 0)
            {
                ds.Tables.Clear();
            }
            string state = cmbState.Text.Trim();
            string id = txtId.Text.Trim();
            string name = txtName.Text.Trim();

            string sql = string.Format("select project_id,project_name,【工程信息】as info from T_project where system_state = '{0}'", state);

            if (id.Length != 0 || name.Length != 0)
```

```csharp
            {
                sql += string.Format(" and project_id like '%{0}%' and project_name like '%{1}%'", id, name);
            }

            db.FillData(sql,ds);
            dgvProject.DataSource = ds.Tables[0];
        }

        private void dgvProject_CellContentClick(object sender, DataGridViewCellEventArgs e)
        {
            if (e.ColumnIndex == 2)
            {
                string id = dgvProject[0, e.RowIndex].Value.ToString();
                frmProjectInfo frm = new frmProjectInfo();
                frm.id = id;
                frm.ShowDialog();
            }
        }
    }
}

/**
 *启动类
 **/
using System;
using System.Collections.Generic;
using System.Linq;
using System.Windows.Forms;

namespace ProjectInfo
{
    static class Program
    {
        /// <summary>
        ///应用程序的主入口点。
        /// </summary>
        [STAThread]
        static void Main()
        {
            Application.EnableVisualStyles();
            Application.SetCompatibleTextRenderingDefault(false);
            Application.Run(new frmLogin());
```

        }
    }
}

项目二

```csharp
/**
 *数据库操作公共类
 **/
using System;
using System.Collections.Generic;
using System.Linq;
using System.Text;
using System.Data;
using System.Data.SqlClient;

namespace ConstructionManager
{
    public class DBHelper
    {
        public string str = @"Data Source=localhost\SA;Initial Catalog=ConstructionDB;Integrated Security=True";
        /// <summary>
        ///增删改
        /// </summary>
        /// <param name="sql"></param>
        /// <returns></returns>
        public int Operate(string sql)
        {
            try
            {
                using (SqlConnection con = new SqlConnection(str))
                using (SqlCommand com = new SqlCommand(sql, con))
                {
                    con.Open();
                    return com.ExecuteNonQuery();
                }
            }
            catch (Exception ex)
            {
                return 0;
```

            }
        }
        /// <summary>
        ///查询结果集
        /// </summary>
        /// <param name="sql"></param>
        /// <returns></returns>
        public SqlDataReader GetReader(string sql)
        {
            try
            {
                SqlConnection con = new SqlConnection(str);
                SqlCommand com = new SqlCommand(sql, con);
                con.Open();
                return com.ExecuteReader(CommandBehavior.CloseConnection);
            }
            catch (Exception ex)
            {

                return null;
            }
        }
}

/**
*用户信息管理界面类
**/
using System;
using System.Collections.Generic;
using System.ComponentModel;
using System.Data;
using System.Drawing;
using System.Linq;
using System.Text;
using System.Windows.Forms;
using System.Data;
using System.Data.SqlClient;

namespace ConstructionManager
{
    public partial class frmMain : Form
    {

```csharp
public frmMain()
{
    InitializeComponent();
}
//退出
private void btnExit_Click(object sender, EventArgs e)
{
    Application.Exit();
}


/// <summary>
/// 显示所有用户
/// </summary>
/// <param name="sender"></param>
/// <param name="e"></param>
private void button1_Click(object sender, EventArgs e)
{
    DBHelper db = new DBHelper();
    lvUser_ID.Items.Clear();//清空原记录
    string sql = "select user_id from T_user";
    SqlDataReader dr = db.GetReader(sql);
    while (dr.Read())
    {
        ListViewItem item = new ListViewItem(dr[0].ToString());
        lvUser_ID.Items.Add(item);
    }
    dr.Close();
}
//单击用户ID显示用户信息
private void lvUser_ID_MouseClick(object sender, MouseEventArgs e)
{
    if(lvUser_ID.SelectedItems.Count > 0)
    {
        DBHelper db = new DBHelper();
        //获取所选的用户的ID
        string userID = lvUser_ID.SelectedItems[0].Text;
        string sql = string.Format("select user_name,user_password,dept_name from T_user where user_id = {0}",userID);
        SqlDataReader dr = db.GetReader(sql);
        while (dr.Read())
        {
            txtUser_name.Text = dr[0].ToString();
            txtUser_password.Text = dr[1].ToString();
```

```csharp
            cboDept_name.Text = dr[2].ToString();
        }
        dr.Close();
    }
}
//清空三个文本框
public void ClearUser()
{
    txtUser_name.Clear();
    txtUser_password.Clear();
    cboDept_name.Text = "";
}


//删除
private void btnDel_Click(object sender, EventArgs e)
{

    if (lvUser_ID.SelectedItems.Count > 0)
    {
        DBHelper db = new DBHelper();
        //获取所选的用户的ID
        string userID = lvUser_ID.SelectedItems[0].Text;
        string sql = string.Format("delete from T_user where user_id = {0}", userID);
        db.Operate(sql);
        //刷新 listView
        button1_Click(null, null);
        //清空三个文本框
        ClearUser();
        MessageBox.Show("删除成功", "系统提示", MessageBoxButtons.OK, MessageBoxIcon.Information);

    }
    else
    {
        MessageBox.Show("您还没有选中一个ID", "系统提示", MessageBoxButtons.OK, MessageBoxIcon.Error);
        return;
    }
}

//设置确定和取消按钮状态
public void SetButton(bool en)
{
```

```
        btnOK.Enabled = en;
        btnCancel.Enabled = en;
    }


    //添加
    private void btnAdd_Click(object sender, EventArgs e)
    {
        //点亮确定和取消按钮
        SetButton(true);
    }

    private void btnCancel_Click(object sender, EventArgs e)
    {
        ClearUser();
        SetButton(false);
    }

    private void btnOK_Click(object sender, EventArgs e)
    {
        string userName = txtUser_name.Text.Trim();
        string pwd = txtUser_password.Text;
        string deptName = cboDept_name.Text;
        if (userName.Length == 0)
        {
            MessageBox.Show("请输入用户名","系统提示",MessageBoxButtons.OK,MessageBoxIcon.Error);
            return;
        }
        if (pwd.Length == 0)
        {
            MessageBox.Show("请输入密码","系统提示",MessageBoxButtons.OK,MessageBoxIcon.Error);
            return;
        }

        string sql = string.Format("insert into T_user values('{0}','{1}','{2}')",userName,pwd,deptName);
        DBHelper db = new DBHelper();
        db.Operate(sql);
        button1_Click(null, null);
        MessageBox.Show("添加成功","系统提示",MessageBoxButtons.OK,MessageBoxIcon.Information);
```

}

```csharp
private void btnUpdate_Click(object sender, EventArgs e)
{
    if (lvUser_ID.SelectedItems.Count > 0)
    {
        DBHelper db = new DBHelper();
        //获取所选的用户的ID
        string userID = lvUser_ID.SelectedItems[0].Text;
        string userName = txtUser_name.Text.Trim();
        string pwd = txtUser_password.Text;
        string deptName = cboDept_name.Text;
        if (userName.Length == 0)
        {
            MessageBox.Show("请输入用户名","系统提示",MessageBoxButtons.OK,MessageBoxIcon.Error);
            return;
        }
        if (pwd.Length == 0)
        {
            MessageBox.Show("请输入密码","系统提示",MessageBoxButtons.OK,MessageBoxIcon.Error);
            return;
        }
        string sql = string.Format("update T_user set user_name='{0}',user_password='{1}',dept_name='{2}'",userName,pwd,deptName);
        db.Operate(sql);
        //刷新 listView
        button1_Click(null, null);
        //清空三个文本框
        ClearUser();
        MessageBox.Show("修改成功","系统提示",MessageBoxButtons.OK,MessageBoxIcon.Information);
    }
    else
    {
        MessageBox.Show("您还没有选中一个ID","系统提示",MessageBoxButtons.OK,MessageBoxIcon.Error);
        return;
    }
}
```

```
        }
}

/**
 *启动类
 **/
using System;
using System.Collections.Generic;
using System.Linq;
using System.Windows.Forms;

namespace ConstructionManager
{
    static class Program
    {
        /// <summary>
        ///应用程序的主入口点。
        /// </summary>
        [STAThread]
        static void Main()
        {
            Application.EnableVisualStyles();
            Application.SetCompatibleTextRenderingDefault(false);
            Application.Run(new frmMain());
        }
    }
}
```

项目三

```
/**
 *数据库操作公共类
 **/
using System;
using System.Collections.Generic;
using System.Linq;
using System.Text;
using System.Data;
using System.Data.SqlClient;

namespace OfferInfo
{
```

```csharp
public class DBHelper
{
    public static string str = @"Data Source=localhost\SA;Initial Catalog=ConstructionDB;Integrated Security=True";

    /// <summary>
    /// 增删改
    /// </summary>
    /// <param name="sql"></param>
    /// <returns></returns>
    public int Operate(string sql)
    {
        using(SqlConnection con = new SqlConnection(str))
        using (SqlCommand com = new SqlCommand(sql, con))
        {
            con.Open();
            return com.ExecuteNonQuery();
        }
    }

    /// <summary>
    /// 查询单个值
    /// </summary>
    /// <param name="sql"></param>
    /// <returns></returns>
    public object GetScalar(string sql)
    {
        using (SqlConnection con = new SqlConnection(str))
        using (SqlCommand com = new SqlCommand(sql, con))
        {
            con.Open();
            return com.ExecuteScalar();
        }
    }

    /// <summary>
    /// 查询结果集
    /// </summary>
    /// <param name="sql"></param>
    /// <returns></returns>
    public SqlDataReader GetReader(string sql)
    {
        SqlConnection con = new SqlConnection(str);
        SqlCommand com = new SqlCommand(sql, con);
        {
```

```csharp
                con.Open();
                return com.ExecuteReader(CommandBehavior.CloseConnection);
        }


        /// <summary>
        ///填充数据集
        /// </summary>
        public SqlDataAdapter da;
        public void FillData(string sql, DataSet ds)
        {
            SqlConnection con = new SqlConnection(str);
            da = new SqlDataAdapter(sql,con);
            da.Fill(ds);
        }

    }
}

/**
*添加投标界面类
**/
using System;
using System.Collections.Generic;
using System.ComponentModel;
using System.Data;
using System.Drawing;
using System.Linq;
using System.Text;
using System.Windows.Forms;

namespace OfferInfo
{
    public partial class frmAddOffer : Form
    {
        public string Project_id = "";
        public frmEnterPriseInfo frm = null;
        public frmAddOffer()
        {
            InitializeComponent();
        }

        private void button2_Click(object sender, EventArgs e)
```

```csharp
{
    this.Close();
}

private void frmAddOffer_Load(object sender, EventArgs e)
{
    // TODO:这行代码将数据加载到表"constructionDBDataSet1.T_enterprise"中。您可以根据需要移动或移除它。
    this.t_enterpriseTableAdapter.Fill(this.constructionDBDataSet1.T_enterprise);
}

//新增
private void button1_Click(object sender, EventArgs e)
{
    DBHelper db = new DBHelper();
    string entid = cmbEntId.Text.Trim();
    string budget_price = txtBudget_price.Text.Trim();
    string offer_price = txtOffer_price.Text.Trim();
    string Ration_limite = txtRation_limite.Text.Trim();
    string offer_limite = txtOffer_limite.Text.Trim();

    if (budget_price.Length == 0)
    {
        MessageBox.Show("请输入预算报价", "系统提示", MessageBoxButtons.OK, MessageBoxIcon.Warning);
        return;
    }
    if (offer_price.Length == 0)
    {
        MessageBox.Show("请输入投标报价", "系统提示", MessageBoxButtons.OK, MessageBoxIcon.Warning);
        return;
    }
    if (Ration_limite.Length == 0)
    {
        MessageBox.Show("请输入定额工期", "系统提示", MessageBoxButtons.OK, MessageBoxIcon.Warning);
        return;
    }
    if (offer_limite.Length == 0)
    {
        MessageBox.Show("请输入投标工期", "系统提示", MessageBoxButtons.OK, Message-
```

```csharp
            BoxIcon.Warning);
                return;
            }
            try
            {
                string sql = string.Format("insert into T_offer values('{0}','{1}','{2}','{3}','{4}','{5}')", Project_id, entid, budget_price, offer_price, Ration_limite, offer_limite);
                db.Operate(sql);
                frm.button1_Click(null,null);
                MessageBox.Show("增加成功!","系统提示",MessageBoxButtons.OK,MessageBoxIcon.Information);
                this.Close();
            }
            catch(Exception ex)
            {
                MessageBox.Show("此企业已参与投标,不能重复参与","系统提示",MessageBoxButtons.OK,MessageBoxIcon.Error);
                return;
            }
        }
    }
}

/**
* 投标企业查询类
* */
using System;
using System.Collections.Generic;
using System.ComponentModel;
using System.Data;
using System.Drawing;
using System.Linq;
using System.Text;
using System.Windows.Forms;

namespace OfferInfo
{
    public partial class frmEnterPriseInfo : Form
    {
        DataSet ds = new DataSet();
        DBHelper db = new DBHelper();
        public frmEnterPriseInfo()
```

```csharp
{
    InitializeComponent();
}

private void Form1_Load(object sender, EventArgs e)
{
    // TODO:这行代码将数据加载到表"constructionDBDataSet.T_project"中。您可以根据需要移动或移除它。
    this.t_projectTableAdapter.Fill(this.constructionDBDataSet.T_project);

}

//关闭
private void button3_Click(object sender, EventArgs e)
{
    Application.Exit();
}

//查询
public void button1_Click(object sender, EventArgs e)
{
    if (ds.Tables.Count != 0)
    {
        ds.Tables.Clear();
    }
    string id = cmbProjectName.SelectedValue.ToString();
    string sql = string.Format("select e.ent_id as eid,e.ent_name as ename,'【删除】' as operate from T_enterprise e,T_offer o where e.ent_id = o.ent_id and project_id = '{0}'", id);
    db.FillData(sql, ds);
    dgvEnterPrise.DataSource = ds.Tables[0];

}

private void button2_Click(object sender, EventArgs e)
{
    string id = cmbProjectName.SelectedValue.ToString();
    frmAddOffer frm = new frmAddOffer();
    frm.Project_id = id;
    frm.frm = this;
    frm.ShowDialog();
}
}
}
```

```
/**
 *启动类
 **/
using System;
using System.Collections.Generic;
using System.Linq;
using System.Windows.Forms;

namespace OfferInfo
{
    static class Program
    {
        /// <summary>
        ///应用程序的主入口点。
        /// </summary>
        [STAThread]
        static void Main()
        {
            Application.EnableVisualStyles();
            Application.SetCompatibleTextRenderingDefault(false);
            Application.Run(new frmEnterPriseInfo());
        }
    }
}
```

项目四

```
/**
 *数据库操作公共类
 **/

using System;
using System.Collections.Generic;
using System.Linq;
using System.Text;
using System.Data;
using System.Data.SqlClient;

namespace HarborBureau
{
    public class DBHelper
```

```csharp
{
    public string str = @"Data Source=localhost\SA;Initial Catalog=HarborBureauDB;Integrated Security=True";
    /// <summary>
    ///增删改
    /// </summary>
    /// <param name="sql"></param>
    /// <returns></returns>
    public int Operate(string sql)
    {
        using(SqlConnection con = new SqlConnection(str))
        using (SqlCommand com = new SqlCommand(sql, con))
        {
            con.Open();
            return com.ExecuteNonQuery();
        }
    }
    /// <summary>
    ///查询单个值
    /// </summary>
    /// <param name="sql"></param>
    /// <returns></returns>
    public object GetScalar(string sql)
    {
        using (SqlConnection con = new SqlConnection(str))
        using (SqlCommand com = new SqlCommand(sql, con))
        {
            con.Open();
            return com.ExecuteScalar();
        }
    }
    /// <summary>
    ///查询结果集
    /// </summary>
    /// <param name="sql"></param>
    /// <returns></returns>
    public SqlDataReader GetReader(string sql)
    {
        SqlConnection con = new SqlConnection(str);
        SqlCommand com = new SqlCommand(sql, con);
        {
            con.Open();
            return com.ExecuteReader(CommandBehavior.CloseConnection);
```

```csharp
        }
        /// <summary>
        ///填充数据集
        /// </summary>
        public SqlDataAdapter da;
        public void FillData(string sql, DataSet ds)
        {
            SqlConnection con = new SqlConnection(str);
            da = new SqlDataAdapter(sql,con);
            da.Fill(ds);
        }
    }
}

/**
*新增航线类
**/
using System;
using System.Collections.Generic;
using System.ComponentModel;
using System.Data;
using System.Drawing;
using System.Linq;
using System.Text;
using System.Windows.Forms;

namespace HarborBureau
{
    public partial class frmAddLine : Form
    {
        public frmAddLine()
        {
            InitializeComponent();
        }

        private void button2_Click(object sender, EventArgs e)
        {
            this.Close();
        }
```

```csharp
private void frmAddLine_Load(object sender, EventArgs e)
{
    // TODO:这行代码将数据加载到表"harborBureauDBDataSet.T_port"中。您可以根据需要移动或移除它。
    this.t_portTableAdapter.Fill(this.harborBureauDBDataSet.T_port);

}

//保存
private void button1_Click(object sender, EventArgs e)
{
    DBHelper db = new DBHelper();
    string line_code = txtLine_code.Text.Trim();
    string line_name = txtLine_name.Text.Trim();
    string port_code = cmbPort_code.Text.Trim();
    string dock_unit_name = txtDock_unit_name.Text.Trim();
    string line_type = txtLine_type.Text.Trim();

    if (line_code.Length == 0)
    {
        MessageBox.Show("航线代码不能为空!","系统提示",MessageBoxButtons.OK,MessageBoxIcon.Warning);
        return;
    }
    if (line_name.Length == 0)
    {
        MessageBox.Show("航线名称不能为空!","系统提示",MessageBoxButtons.OK,MessageBoxIcon.Warning);
        return;
    }
    if (port_code.Length == 0)
    {
        MessageBox.Show("港口代码不能为空!","系统提示",MessageBoxButtons.OK,MessageBoxIcon.Warning);
        return;
    }
    if (dock_unit_name.Length == 0)
    {
        MessageBox.Show("码头名称不能为空!","系统提示",MessageBoxButtons.OK,MessageBoxIcon.Warning);
        return;
    }
    if (line_type.Length == 0)
```

```csharp
            {
                MessageBox.Show("航线类型不能为空!","系统提示",MessageBoxButtons.OK,MessageBoxIcon.Warning);
                return;
            }
            try
            {
                string sql = string.Format("select count(*) from T_line where line_code = '{0}'", line_code);
                int n = Convert.ToInt32(db.GetScalar(sql));
                if (n > 0)
                {
                    MessageBox.Show("航线代码已存在!","系统提示",MessageBoxButtons.OK,MessageBoxIcon.Warning);
                    return;
                }

                sql = string.Format("insert into T_line values('{0}','{1}','{2}','{3}','{4}')", line_code, line_name, port_code, dock_unit_name, line_type);
                db.Operate(sql);
                MessageBox.Show("添加成功!","系统提示",MessageBoxButtons.OK,MessageBoxIcon.Information);
                this.Close();
            }
            catch (Exception ex)
            {
                MessageBox.Show("操作失败!","系统提示",MessageBoxButtons.OK,MessageBoxIcon.Error);
                return;
            }
        }
    }

/**
 * 航线管理类
 **/
using System;
using System.Collections.Generic;
using System.ComponentModel;
using System.Data;
```

```csharp
using System.Drawing;
using System.Linq;
using System.Text;
using System.Windows.Forms;

namespace HarborBureau
{
    public partial class frmLine_manager : Form
    {
        DataSet ds = new DataSet();
        DBHelper db = new DBHelper();
        public frmLine_manager()
        {
            InitializeComponent();
        }

        private void button1_Click(object sender, EventArgs e)
        {
            if (ds.Tables.Count != 0)
            {
                ds.Tables[0].Clear();
            }
            string line_code = cmbLine_code.Text.Trim();
            string line_name = cmbLine_name.Text.Trim();
            string port_code = cmbPort_code.Text.Trim();
            string dock_unit_name = cmbDock_unit_name.Text.Trim();
            string line_type = cmbLine_type.Text.Trim();
            string sql = "select * from T_line where 1=1 ";
            if (line_code.Length != 0)
            {
                sql += string.Format(" and Line_code = '{0}'", line_code);
            }
            if (line_name.Length != 0)
            {
                sql += string.Format(" and Line_name = '{0}'", line_name);
            }
            if (port_code.Length != 0)
            {
                sql += string.Format(" and Port_code = '{0}'", port_code);
            }
            if (dock_unit_name.Length != 0)
            {
                sql += string.Format(" and Dock_unit_name = '{0}'", dock_unit_name);
```

```
            }
            if (line_type.Length != 0)
            {
                sql += string.Format(" and Line_type = '{0}'", line_type);
            }
            try
            {
                db.FillData(sql, ds);
                dgvLine_info.DataSource = ds.Tables[0];
                if (ds.Tables[0].Rows.Count == 0)
                {
                    MessageBox.Show("对不起,没有符合条件的记录", "系统提示", MessageBoxButtons.OK, MessageBoxIcon.Error);
                }
            }
            catch (Exception ex)
            {
                MessageBox.Show("操作失败!", "系统提示", MessageBoxButtons.OK, MessageBoxIcon.Error);
                return;
            }
        }

        private void button2_Click(object sender, EventArgs e)
        {
            frmAddLine frm = new frmAddLine();
            frm.ShowDialog();
        }
    }
}

/**
 * 启动类
 **/
using System;
using System.Collections.Generic;
using System.Linq;
using System.Windows.Forms;

namespace HarborBureau
{
```

— 124 —

```
static class Program
{
    /// <summary>
    /// 应用程序的主入口点。
    /// </summary>
    [STAThread]
    static void Main()
    {
        Application.EnableVisualStyles();
        Application.SetCompatibleTextRenderingDefault(false);
        Application.Run(new frmLine_manager());
    }
}
```

项目五

```
/**
 *数据库操作公共类
 **/
using System;
using System.Collections.Generic;
using System.Linq;
using System.Text;
using System.Data;
using System.Data.SqlClient;

namespace HarborBureauManager
{
    public class DBHelper
    {
        public static string str = @"Data Source=localhost\SA;Initial Catalog=HarborBureauDB;Integrated Security=True";
        /// <summary>
        /// 增删改
        /// </summary>
        /// <param name="sql"></param>
        /// <returns></returns>
        public int Operate(string sql)
        {
            using(SqlConnection con = new SqlConnection(str))
            using (SqlCommand com = new SqlCommand(sql, con))
```

```csharp
        {
            con.Open();
            return com.ExecuteNonQuery();
        }
    }
    /// <summary>
    ///查询单个值
    /// </summary>
    /// <param name="sql"></param>
    /// <returns></returns>
    public object GetScalar(string sql)
    {
        using (SqlConnection con = new SqlConnection(str))
        using (SqlCommand com = new SqlCommand(sql, con))
        {
            con.Open();
            return com.ExecuteScalar();
        }
    }
    /// <summary>
    ///查询结果集
    /// </summary>
    /// <param name="sql"></param>
    /// <returns></returns>
    public SqlDataReader GetReader(string sql)
    {
        SqlConnection con = new SqlConnection(str);
        SqlCommand com = new SqlCommand(sql, con);
        {
            con.Open();
            return com.ExecuteReader(CommandBehavior.CloseConnection);
        }
    }

    /// <summary>
    ///填充数据集
    /// </summary>
    public SqlDataAdapter da;
    public void FillData(string sql, DataSet ds)
    {
        SqlConnection con = new SqlConnection(str);
        da = new SqlDataAdapter(sql, con);
        da.Fill(ds);
```

            }

        }
}

/**
 *保安员证件管理类
 **/
using System;
using System.Collections.Generic;
using System.ComponentModel;
using System.Data;
using System.Drawing;
using System.Linq;
using System.Text;
using System.Windows.Forms;

namespace HarborBureauManager
{
    public partial class frmFacility_SecurityManager : Form
    {
        DataSet ds = new DataSet();
        DBHelper db = new DBHelper();
        public frmFacility_SecurityManager()
        {
            InitializeComponent();
        }

        private void dgvFacility_Security_CellContentClick(object sender, DataGridViewCellEventArgs e)
        {

        }

        private void frmFacility_SecurityManager_Load(object sender, EventArgs e)
        {
            // TODO:这行代码将数据加载到表"harborBureauDBDataSet.T_facility_security"中。您可以根据需要移动或移除它。
            //this.t_facility_securityTableAdapter.Fill(this.harborBureauDBDataSet.T_facility_security);
            ShowInfo();
        }

        public void ShowInfo()

```csharp
        {
            if (ds.Tables.Count != 0)
            {
                ds.Tables[0].Clear();
            }
            string sql = "select * from T_facility_security";
            db.FillData(sql, ds);
            dgvFacility_Security.DataSource = ds.Tables[0];
        }

        private void button2_Click(object sender, EventArgs e)
        {
            this.Close();
        }

        private void button1_Click(object sender, EventArgs e)
        {
            if (dgvFacility_Security.SelectedRows.Count == 0)
            {
                MessageBox.Show("请选择要修改的证件信息", "系统提示", MessageBoxButtons.OK, MessageBoxIcon.Warning);
                return;
            }
            else
            {
                string no = dgvFacility_Security.SelectedRows[0].Cells[0].Value.ToString();
                string name = dgvFacility_Security.SelectedRows[0].Cells[1].Value.ToString();
                string unit_name = dgvFacility_Security.SelectedRows[0].Cells[2].Value.ToString();
                string date = dgvFacility_Security.SelectedRows[0].Cells[3].Value.ToString();
                string type = dgvFacility_Security.SelectedRows[0].Cells[4].Value.ToString();
                frmModify frm = new frmModify();
                frm.no = no;
                frm.name = name;
                frm.unit_name = unit_name;
                frm.date = date;
                frm.type = type;
                frm.frm = this;
                frm.ShowDialog();
            }
        }
    }
}
```

```csharp
/**
 *登录类
 **/
using System;
using System.Collections.Generic;
using System.ComponentModel;
using System.Data;
using System.Drawing;
using System.Linq;
using System.Text;
using System.Windows.Forms;

namespace HarborBureauManager
{
    public partial class frmLogin : Form
    {
        public frmLogin()
        {
            InitializeComponent();
        }

        private void button2_Click(object sender, EventArgs e)
        {
            Application.Exit();
        }

        private void button1_Click(object sender, EventArgs e)
        {
            DBHelper db = new DBHelper();
            string user = txtUser.Text.Trim();
            string pwd = txtPwd.Text;
            if (user.Length == 0)
            {
                MessageBox.Show("请输入用户名", "系统提示", MessageBoxButtons.OK, MessageBoxIcon.Warning);
                txtUser.Focus();
                return;
            }
            if (pwd.Length == 0)
            {
                MessageBox.Show("请输入密码", "系统提示", MessageBoxButtons.OK, MessageBoxIcon.Warning);
                txtPwd.Focus();
```

```csharp
                return;
            }
            string sql = string.Format("select M_password from T_manager where M_user_name ='{0}'", user);
            string realPwd = Convert.ToString(db.GetScalar(sql));
            if (pwd == realPwd)
            {
                frmFacility_SecurityManager frm = new frmFacility_SecurityManager();
                frm.Show();
                this.Hide();
            }
            else
            {
                MessageBox.Show("用户名或密码有误!", "系统提示", MessageBoxButtons.OK, MessageBoxIcon.Error);
                return;
            }
        }
    }
}

/**
 *保安员证件修改类
 **/
using System;
using System.Collections.Generic;
using System.ComponentModel;
using System.Data;
using System.Drawing;
using System.Linq;
using System.Text;
using System.Windows.Forms;

namespace HarborBureauManager
{
    public partial class frmModify : Form
    {
        public string no = "";
        public string name = "";
        public string unit_name = "";
        public string date = "";
        public string type = "";
```

```csharp
        public frmFacility_SecurityManager frm;
        public frmModify()
        {
            InitializeComponent();
        }

        private void frmModify_Load(object sender, EventArgs e)
        {
            txtNO.Text = no;
            txtName.Text = name;
            txtUnit.Text = unit_name;
            txtDate.Text = date;
            cmbType.Text = type;
        }

        private void button2_Click(object sender, EventArgs e)
        {
            this.Close();
        }

        private void button1_Click(object sender, EventArgs e)
        {
            string newno = txtNO.Text.Trim();
            string name = txtName.Text.Trim();
            string unit = txtUnit.Text.Trim();
            string date = txtDate.Text.Trim();
            string type = cmbType.Text.Trim();
            string sql = string.Format("update T_facility_security set certificate_no = '{0}',name = '{1}',unit_name = '{2}',effective_date = '{3}',certificate_type = '{4}' where Certificate_no = '{5}'",newno,name,unit,date,type,no);
            DBHelper db = new DBHelper();
            db.Operate(sql);
            frm.ShowInfo();
            MessageBox.Show("修改成功!");
            this.Close();
        }
    }
}

/**
 *启动类
 **/
using System;
```

```csharp
using System.Collections.Generic;
using System.Linq;
using System.Windows.Forms;

namespace HarborBureauManager
{
    static class Program
    {
        /// <summary>
        /// 应用程序的主入口点。
        /// </summary>
        [STAThread]
        static void Main()
        {
            Application.EnableVisualStyles();
            Application.SetCompatibleTextRenderingDefault(false);
            Application.Run(new frmLogin());
        }
    }
}
```

项目六

```csharp
/**
 *数据库操作公共类
 **/
using System;
using System.Collections.Generic;
using System.Linq;
using System.Text;
using System.Data;
using System.Data.SqlClient;

namespace HarborBureauManager
{
    public class DBHelper
    {
        public static string str = @"Data Source=localhost\SA;Initial Catalog=HarborBureauDB;Integrated Security=True";
        /// <summary>
        /// 增删改
        /// </summary>
```

/// <param name="sql"></param>
/// <returns></returns>
public int Operate(string sql)
{
    using(SqlConnection con = new SqlConnection(str))
    using (SqlCommand com = new SqlCommand(sql, con))
    {
        con.Open();
        return com.ExecuteNonQuery();
    }
}
/// <summary>
///查询单个值
/// </summary>
/// <param name="sql"></param>
/// <returns></returns>
public object GetScalar(string sql)
{
    using (SqlConnection con = new SqlConnection(str))
    using (SqlCommand com = new SqlCommand(sql, con))
    {
        con.Open();
        return com.ExecuteScalar();
    }
}
/// <summary>
///查询结果集
/// </summary>
/// <param name="sql"></param>
/// <returns></returns>
public SqlDataReader GetReader(string sql)
{
    SqlConnection con = new SqlConnection(str);
    SqlCommand com = new SqlCommand(sql,con);
    con.Open();
    return com.ExecuteReader(CommandBehavior.CloseConnection);
}

/// <summary>
///填弃数据集
/// </summary>
public SqlDataAdapter da;
public void FillData(string sql, DataSet ds)

```
        {
            SqlConnection con = new SqlConnection(str);
            da = new SqlDataAdapter(sql,con);
            da.Fill(ds);
        }


    }
}

/**
 *增加船货信息类
 **/
using System;
using System.Collections.Generic;
using System.ComponentModel;
using System.Data;
using System.Drawing;
using System.Linq;
using System.Text;
using System.Windows.Forms;

namespace HarborBureauManager
{
    public partial class frmAddDeclare : Form
    {
        public frmDeclareManager frm;
        public frmAddDeclare()
        {
            InitializeComponent();
        }

        private void button2_Click(object sender, EventArgs e)
        {
            this.Close();
        }

        private void button1_Click(object sender, EventArgs e)
        {
            DBHelper db = new DBHelper();
            string no = txtNO.Text.Trim();
            string name = txtName.Text.Trim();
            string location = txtLocation.Text.Trim();
```

```csharp
            string cargo = txtCargo.Text.Trim();
            string qty = txtQty.Text.Trim();

            if (no.Length == 0)
            {
                MessageBox.Show("请输入船货申请编号","系统提示",MessageBoxButtons.OK,MessageBoxIcon.Warning);
                return;
            }
            if (name.Length == 0)
            {
                MessageBox.Show("请输入船泊名称","系统提示",MessageBoxButtons.OK,MessageBoxIcon.Warning);
                return;
            }
            if (location.Length == 0)
            {
                MessageBox.Show("请输入泊位位置","系统提示",MessageBoxButtons.OK,MessageBoxIcon.Warning);
                return;
            }
            if (cargo.Length == 0)
            {
                MessageBox.Show("请输入货物名称","系统提示",MessageBoxButtons.OK,MessageBoxIcon.Warning);
                return;
            }
            if (qty.Length == 0)
            {
                MessageBox.Show("请输入集装箱数量","系统提示",MessageBoxButtons.OK,MessageBoxIcon.Warning);
                return;
            }
            try
            {
                string sql = string.Format("select count(*) from T_cargo_declare where declare_no ='{0}'", no);
                int n = Convert.ToInt32(db.GetScalar(sql));
                if (n > 0)
                {
                    MessageBox.Show("此船货申请编号已存在!","系统提示",MessageBoxButtons.OK,MessageBoxIcon.Error);
                    return;
```

```
            }
                    sql = string.Format("insert into T_cargo_declare values('{0}','{1}','{2}','{3}','{4}')", no, name, location, cargo, qty);
                    db.Operate(sql);
                    frm.ShowDeclare();
                    MessageBox.Show("添加成功!","系统提示",MessageBoxButtons.OK,MessageBoxIcon.Information);
                    this.Close();
            }
            catch(Exception ex)
            {
                    MessageBox.Show("未知错误,添加失败!","系统提示",MessageBoxButtons.OK,MessageBoxIcon.Error);
                    return;
            }
        }
}

/**
 * 船货信息管理类
 */
using System;
using System.Collections.Generic;
using System.ComponentModel;
using System.Data;
using System.Drawing;
using System.Linq;
using System.Text;
using System.Windows.Forms;

namespace HarborBureauManager
{
    public partial class frmDeclareManager : Form
    {
        DBHelper db = new DBHelper();
        DataSet ds = new DataSet();
        public frmDeclareManager()
        {
            InitializeComponent();
        }
```

```csharp
//关闭
private void button3_Click(object sender, EventArgs e)
{
    Application.Exit();
}

private void frmDeclareManager_Load(object sender, EventArgs e)
{
    // TODO:这行代码将数据加载到表"harborBureauDBDataSet.T_cargo_declare"中。您可以根据需要移动或移除它。
    //this.t_cargo_declareTableAdapter.Fill(this.harborBureauDBDataSet.T_cargo_declare);
    ShowDeclare();
}

public void ShowDeclare()
{
    if (ds.Tables.Count != 0)
    {
        ds.Tables[0].Clear();
    }
    string sql = "select * from T_cargo_declare";
    db.FillData(sql, ds);
    dgvDeclare.DataSource = ds.Tables[0];
}

//删除
private void button1_Click(object sender, EventArgs e)
{
    if (dgvDeclare.SelectedRows.Count != 0)
    {
        DialogResult res;
        res = MessageBox.Show("您确认删除吗?", "系统提示", MessageBoxButtons.OKCancel, MessageBoxIcon.Question);
        if (res == DialogResult.OK)
        {
            //获得船货申请编号
            string no = dgvDeclare.SelectedRows[0].Cells[0].Value.ToString();
            string sql = string.Format("delete from T_cargo_declare where Declare_no = '{0}'", no);
            db.Operate(sql);
            ShowDeclare();
            MessageBox.Show("删除成功!", "系统提示", MessageBoxButtons.OK, MessageBoxIcon.Information);
```

```csharp
            }
        }
        else
        {
            MessageBox.Show("请选中一条记录","系统提示",MessageBoxButtons.OK,MessageBoxIcon.Error);
            return;
        }
    }

    private void button2_Click(object sender, EventArgs e)
    {
        frmAddDeclare frm = new frmAddDeclare();
        frm.frm = this;
        frm.ShowDialog();
    }
}

/**
 *启动类
 **/
using System;
using System.Collections.Generic;
using System.Linq;
using System.Windows.Forms;

namespace HarborBureauManager
{
    static class Program
    {
        /// <summary>
        /// 应用程序的主入口点。
        /// </summary>
        [STAThread]
        static void Main()
        {
            Application.EnableVisualStyles();
            Application.SetCompatibleTextRenderingDefault(false);
            Application.Run(new frmDeclareManager());
        }
    }
}
```

项目七

```csharp
/**
 *数据库操作公共类
 **/
using System;
using System.Collections.Generic;
using System.Linq;
using System.Text;
using System.Data;
using System.Data.SqlClient;

namespace ProductManager
{
    public class DBHelper
    {
        public static string str = @"Data Source=localhost\SA;Initial Catalog=ProductDB;Integrated Security=True";
        /// <summary>
        ///填充数据集
        /// </summary>
        public SqlDataAdapter da;
        public void FillData(string sql, DataSet ds)
        {
            SqlConnection con = new SqlConnection(str);
            da = new SqlDataAdapter(sql, con);
            da.Fill(ds);
        }
    }
}

/**
 *产品管理类
 **/
using System;
using System.Collections.Generic;
using System.ComponentModel;
using System.Data;
using System.Drawing;
using System.Linq;
using System.Text;
```

```csharp
using System.Windows.Forms;

namespace ProductManager
{
    public partial class frmMain : Form
    {
        DBHelper db = new DBHelper();
        DataSet ds = new DataSet();
        public frmMain()
        {
            InitializeComponent();
        }

        private void frmMain_Load(object sender, EventArgs e)
        {
            // TODO:这行代码将数据加载到表"productDBDataSet1.T_category"中。您可以根据需要移动或移除它。
            this.t_categoryTableAdapter.Fill(this.productDBDataSet1.T_category);
            ShowAllProduct();

        }

        //全部产品
        private void button2_Click(object sender, EventArgs e)
        {
            ShowAllProduct();
        }

        private void ShowAllProduct()
        {
            if (ds.Tables.Count != 0)
            {
                ds.Tables[0].Clear();
            }
            string sql = "select product_name,price,remark,register_date from T_product";
            db.FillData(sql, ds);
            dgvProduct.DataSource = ds.Tables[0];
        }

        private void button1_Click(object sender, EventArgs e)
        {
            if (ds.Tables.Count != 0)
            {
```

```
                    ds.Tables[0].Clear();
                }
                int cid = Convert.ToInt32(cmbCategory.SelectedValue);
                string sql = string.Format("select product_name,price,remark,register_date from T_product where category_id = {0}",cid);
                db.FillData(sql,ds);
                dgvProduct.DataSource = ds.Tables[0];
            }
        }
    }
```

/**
*启动类
**/
using System;
using System.Collections.Generic;
using System.Linq;
using System.Windows.Forms;

namespace ProductManager
{
    static class Program
    {
        /// <summary>
        ///应用程序的主入口点。
        /// </summary>
        [STAThread]
        static void Main()
        {
            Application.EnableVisualStyles();
            Application.SetCompatibleTextRenderingDefault(false);
            Application.Run(new frmMain());
        }
    }
}

项目八

/**
*数据库操作公共类
**/
using System;

```csharp
using System.Collections.Generic;
using System.Linq;
using System.Text;
using System.Data;
using System.Data.SqlClient;

namespace ProductManager
{
    public class DBHepler
    {
        public static string str = @"Data Source=localhost\SA;Initial Catalog=ProductDB;Integrated Security=True";
        /// <summary>
        ///增删改
        /// </summary>
        /// <param name="sql"></param>
        /// <returns></returns>
        public int Operate(string sql)
        {
            using(SqlConnection con = new SqlConnection(str))
            using (SqlCommand com = new SqlCommand(sql, con))
            {
                con.Open();
                return com.ExecuteNonQuery();
            }
        }
        /// <summary>
        ///查询单个值
        /// </summary>
        /// <param name="sql"></param>
        /// <returns></returns>
        public object GetScalar(string sql)
        {
            using (SqlConnection con = new SqlConnection(str))
            using (SqlCommand com = new SqlCommand(sql, con))
            {
                con.Open();
                return com.ExecuteScalar();
            }
        }

        /// <summary>
        ///填充数据集
```

```
        /// </summary>
        public SqlDataAdapter da;
        public void FillData(string sql, DataSet ds)
        {
            SqlConnection con = new SqlConnection(str);
            da = new SqlDataAdapter(sql,con);
            da.Fill(ds);
        }
    }
}

/**
 *产品添加管理类
 **/
using System;
using System.Collections.Generic;
using System.ComponentModel;
using System.Data;
using System.Drawing;
using System.Linq;
using System.Text;
using System.Windows.Forms;

namespace ProductManager
{
    public partial class frmMain : Form
    {
        DBHepler db = new DBHepler();
        DataSet ds = new DataSet();
        public frmMain()
        {
            InitializeComponent();
        }

        private void frmMain_Load(object sender, EventArgs e)
        {
            ShowProduct();
        }

        private void ShowProduct()
        {
            if (ds.Tables.Count != 0)
            {
```

```csharp
            ds.Tables[0].Clear();
        }
        string sql = "select Product_id,category_name,product_name,price,remark,p.register_date from T_category c,T_product p where c.category_id = p.category_id";
        db.FillData(sql, ds);
        dgvProduct.DataSource = ds.Tables[0];
    }

    private void button1_Click(object sender, EventArgs e)
    {
        string category_id = txtCategoryId.Text.Trim();
        string product_name = txtProduct_Name.Text.Trim();
        string price = txtPrice.Text.Trim();
        string remark = txtRemark.Text.Trim();

        if (category_id.Length == 0)
        {
            MessageBox.Show("请输入类别编号", "系统提示", MessageBoxButtons.OK, MessageBoxIcon.Error);
            return;
        }
        try
        {
            int cid = Convert.ToInt32(category_id);
        }catch(Exception ex)
        {
            MessageBox.Show("类别编号必须为数字", "系统提示", MessageBoxButtons.OK, MessageBoxIcon.Error);
            return;
        }

        if (product_name.Length == 0)
        {
            MessageBox.Show("请输入产品名称", "系统提示", MessageBoxButtons.OK, MessageBoxIcon.Error);
            return;
        }

        if (price.Length == 0)
        {
            MessageBox.Show("请输入产品价格", "系统提示", MessageBoxButtons.OK, MessageBoxIcon.Error);
            return;
```

```csharp
            }
            try
            {
                double dprice = Convert.ToDouble(price);
                if (dprice < 1 || dprice > 2000)
                {
                    MessageBox.Show("价格必须在1-2000之间","系统提示",MessageBoxButtons.OK,MessageBoxIcon.Error);
                    return;
                }
            }
            catch (Exception ex)
            {
                MessageBox.Show("价格必须为数字","系统提示",MessageBoxButtons.OK,MessageBoxIcon.Error);
                return;
            }

            //验证类别编号是否存在
            string sql = string.Format("select count(*) from T_category where category_id={0}",category_id);
            int n = Convert.ToInt32(db.GetScalar(sql));
            if (n < 1)
            {
                MessageBox.Show("此类别编号不存在","系统提示",MessageBoxButtons.OK,MessageBoxIcon.Error);
                return;
            }

            //设置产品编号
            //先找到最大的产品编号
            sql = "select max(product_id) from T_product";
            int maxProduct_id = Convert.ToInt32(db.GetScalar(sql));
            int product_id = maxProduct_id + 1;  //新产品的编号

            //添加
            sql = string.Format("insert into T_product values({0},{1},'{2}',{3},'{4}',getdate())",product_id,category_id,product_name,price,remark);
            db.Operate(sql);
            MessageBox.Show("添加成功!","系统提示",MessageBoxButtons.OK,MessageBoxIcon.Information);
            txtCategoryId.Text = "";
```

```csharp
            txtProduct_Name.Text = "";
            txtPrice.Text = "";
            txtRemark.Text = "";
            ShowProduct();
        }
    }
}

/**
 *启动类
 **/
using System;
using System.Collections.Generic;
using System.Linq;
using System.Windows.Forms;

namespace ProductManager
{
    static class Program
    {
        /// <summary>
        ///应用程序的主入口点。
        /// </summary>
        [STAThread]
        static void Main()
        {
            Application.EnableVisualStyles();
            Application.SetCompatibleTextRenderingDefault(false);
            Application.Run(new frmMain());
        }
    }
}
```

项目九

```csharp
/**
 *数据库操作公共类
 **/
using System;
using System.Collections.Generic;
using System.Linq;
```

```csharp
using System.Text;
using System.Data;
using System.Data.SqlClient;

namespace CardManager
{
    public class DBHelper
    {
        public string str = @"Data Source=localhost\SA;Initial Catalog=CardDB;Integrated Security=True";
        /// <summary>
        /// 增删改
        /// </summary>
        /// <param name="sql"></param>
        /// <returns></returns>
        public int Operate(string sql)
        {
            using(SqlConnection con = new SqlConnection(str))
            using (SqlCommand com = new SqlCommand(sql, con))
            {
                con.Open();
                return com.ExecuteNonQuery();
            }
        }

        /// <summary>
        /// 返回单个值
        /// </summary>
        /// <param name="sql"></param>
        /// <returns></returns>
        public object GetScalar(string sql)
        {
            using (SqlConnection con = new SqlConnection(str))
            using (SqlCommand com = new SqlCommand(sql, con))
            {
                con.Open();
                return com.ExecuteScalar();
            }
        }

        /// <summary>
        /// 返回结果集
        /// </summary>
```

```csharp
        /// <param name="sql"></param>
        /// <returns></returns>
        public SqlDataReader GetReader(string sql)
        {
            SqlConnection con = new SqlConnection(str);
            SqlCommand com = new SqlCommand(sql, con);
            {
                con.Open();
                return com.ExecuteReader(CommandBehavior.CloseConnection);
            }
        }


        /// <summary>
        ///填充数据集
        /// </summary>
        public SqlDataAdapter da;
        public void FillData(string sql, DataSet ds)
        {
            SqlConnection con = new SqlConnection(str);
            da = new SqlDataAdapter(sql,con);
            da.Fill(ds);
        }



    }
}

/**
*饭卡添加管理类
**/
using System;
using System.Collections.Generic;
using System.ComponentModel;
using System.Data;
using System.Drawing;
using System.Linq;
using System.Text;
using System.Windows.Forms;

namespace CardManager
{
    public partial class frmMain : Form
    {
```

```csharp
DataSet ds = new DataSet();
DBHelper db = new DBHelper();
public frmMain()
{
    InitializeComponent();
}
//窗体加载时,显示所有的饭卡信息
private void frmMain_Load(object sender, EventArgs e)
{
    ShowCard();
}
//显示饭卡信息的方法
public void ShowCard()
{
    //先清空原有记录
    if (ds.Tables.Count != 0)
    {
        ds.Tables[0].Clear();
    }
    string sql = "select * from T_card";
    db.FillData(sql, ds);
    dgvCard.DataSource = ds.Tables[0];
}


//添加饭卡
private void button1_Click(object sender, EventArgs e)
{
    string studentId = txtStudent_id.Text.Trim();
    string name = txtStudent_name.Text.Trim();
    string money = txtMoney.Text.Trim();
    if (studentId.Length == 0)
    {
        MessageBox.Show("请输入学号", "系统提示", MessageBoxButtons.OK, MessageBoxIcon.Error);
        txtStudent_id.Focus();
        return;
    }
    try
    {
        int id = Convert.ToInt32(studentId);
    }
    catch (Exception ex)
    {
```

```csharp
            MessageBox.Show("学号必须为数字","系统提示",MessageBoxButtons.OK,MessageBoxIcon.Error);
            txtStudent_id.Select();
            return;
        }

        if (name.Length == 0)
        {
            MessageBox.Show("请输入姓名","系统提示",MessageBoxButtons.OK,MessageBoxIcon.Error);
            txtStudent_name.Focus();
            return;
        }
        if (money.Length == 0)
        {
            MessageBox.Show("请输入金额","系统提示",MessageBoxButtons.OK,MessageBoxIcon.Error);
            txtMoney.Focus();
            return;
        }
        try
        {
            int moneyInt = Convert.ToInt32(money);
            if (moneyInt < 50 || moneyInt > 500)
            {
                MessageBox.Show("金额必须在50-500之间","系统提示",MessageBoxButtons.OK,MessageBoxIcon.Error);
                txtMoney.Select();
                return;
            }
        }
        catch (Exception ex)
        {
            MessageBox.Show("金额必须为数字","系统提示",MessageBoxButtons.OK,MessageBoxIcon.Error);
            txtStudent_id.Select();
            return;
        }

        //添加到数据库
        //处理编号

        string sql = "select max(Card_id) from T_card";
```

```csharp
            int maxId = Convert.ToInt32(db.GetScalar(sql));
            int Card_id = maxId + 1;//编号为最大的编号+1
            sql = string.Format("insert into T_card values({0},{1},'{2}',{3},default)", Card_id, studentId, name, money);
            db.Operate(sql);
            ShowCard();//刷新显示
            MessageBox.Show("恭喜你,添加成功!", "系统提示", MessageBoxButtons.OK, MessageBoxIcon.Information);
            txtStudent_id.Text = "";
            txtStudent_name.Text = "";
            txtMoney.Text = "";
        }
    }
}

/**
 * 启动类
 * */
using System;
using System.Collections.Generic;
using System.Linq;
using System.Windows.Forms;

namespace CardManager
{
    static class Program
    {
        /// <summary>
        /// 应用程序的主入口点。
        /// </summary>
        [STAThread]
        static void Main()
        {
            Application.EnableVisualStyles();
            Application.SetCompatibleTextRenderingDefault(false);
            Application.Run(new frmMain());
        }
    }
}
```

项目十

/**

* 实施备案项目界面类
* */
using System;
using System.Collections.Generic;
using System.ComponentModel;
using System.Data;
using System.Drawing;
using System.Linq;
using System.Text;
using System.Windows.Forms;

using BLL;
using Model;

namespace WindowsFormsApplication1
{
    public partial class Form1 : Form
    {
        recordManager recordManager = new recordManager();

        public Form1()
        {
            InitializeComponent();
            this.MaximizeBox = false;
        }

        private void 退出ToolStripMenuItem_Click(object sender, EventArgs e)
        {
            this.Dispose();
        }

        private void Form1_Load(object sender, EventArgs e)
        {
            cmbName.DataSource = recordManager.GetAllrecord();
            cmbName.DisplayMember = "Proj_name";
            cmbName.ValueMember = "guid";
        }

        private void cmbName_SelectionChangeCommitted(object sender, EventArgs e)
        {
            record record = recordManager.GetrecordById(cmbName.SelectedValue.ToString());
            txtGuid.Text = record.Guid;

```csharp
            txtName.Text = record.Proj_name;
            txtMake_unit.Text = record.Make_unit;
            dtpTime.Value = record.Time;

            cmbResult.Text = "";
            cmbResult.Items.Clear();
            cmbResult.SelectedText = record.Result;
        }

        private void btnAdd_Click(object sender, EventArgs e)
        {
            if (txtGuid.Text == "")
            {
                MessageBox.Show("请输入项目编号!","错误提示",MessageBoxButtons.OK,MessageBoxIcon.Stop);
                return;
            }
            if (txtName.Text == "")
            {
                MessageBox.Show("请输入项目名称!","错误提示",MessageBoxButtons.OK,MessageBoxIcon.Stop);
                return;
            }
            if ( txtMake_unit.Text == "")
            {
                MessageBox.Show("请输入申请单位!","错误提示",MessageBoxButtons.OK,MessageBoxIcon.Stop);
                return;
            }
            if (cmbResult.Text != "成功" && cmbResult.Text != "失败" && cmbResult.Text != "待审")
            {
                MessageBox.Show("请选择正确的处理结果!","错误提示",MessageBoxButtons.OK,MessageBoxIcon.Stop);
                return;
            }
            record record = new record();
            record.Guid = txtGuid.Text;
            record.Proj_name=txtName.Text ;
            record.Make_unit=txtMake_unit.Text;
            record.Time=dtpTime.Value;
            record.Result = cmbResult.Text;
```

```csharp
            if (recordManager.Addrecord(record))
            {
                MessageBox.Show("添加成功","提示",MessageBoxButtons.OK,MessageBoxIcon.Information);
                cmbName.DataSource = null;
                cmbName.DataSource = recordManager.GetAllrecord();
                cmbName.DisplayMember = "Proj_name";
                cmbName.ValueMember = "guid";
                return;
            }
            else
                MessageBox.Show("添加失败！\n编号或项目名重复","错误提示",MessageBoxButtons.RetryCancel,MessageBoxIcon.Stop);
        }


        private void 查询ToolStripMenuItem_Click(object sender, EventArgs e)
        {
            btnAdd.Visible = false;
            label9.Visible = true;
            cmbName.Visible = true;
        }

        private void 增加ToolStripMenuItem1_Click(object sender, EventArgs e)
        {
            btnAdd.Visible = true;
            label9.Visible = false;
            cmbName.Visible = false;
            txtName.Text = "";
            txtMake_unit.Text = "";
            txtGuid.Text = "";

            dtpTime.Value = DateTime.Now;
            cmbResult.Text = "";
            cmbResult.Items.Clear();
            cmbResult.Items.AddRange(new object[] { "成功","失败","待审" });
        }
    }
}

/**
 * 启动类
```

* */
using System;
using System.Collections.Generic;
using System.Linq;
using System.Windows.Forms;

namespace WindowsFormsApplication1
{
    static class Program
    {
        /// <summary>
        ///应用程序的主入口点。
        /// </summary>
        [STAThread]
        static void Main()
        {
            Application.EnableVisualStyles();
            Application.SetCompatibleTextRenderingDefault(false);
            Application.Run(new Form1());
        }
    }
}

/**
 *数据库操作类
 * */
using System;
using System.Collections.Generic;
using System.Linq;
using System.Text;

using Model;
using System.Configuration;
using System.Data.SqlClient;
using System.Data;

namespace DAL
{
    public class recordServer
    {

        #region Private Members
        //从配置文件中读取数据库连接字符串

```csharp
            private readonly string connString = ConfigurationManager.ConnectionStrings["ConnectionString"].ToString();
            private readonly string dboOwner = ConfigurationManager.ConnectionStrings["DataBaseOwner"].ToString();
            #endregion



            /// <summary>
            ///获取所有
            /// </summary>
            /// <param name="proj_name"></param>
            /// <returns></returns>
            public IList<record> GetAllrecord()
            {
                IList<record> records = new List<record>();
                using (SqlConnection conn = new SqlConnection(connString))
                {
                    string sql = "select * from T_record";
                    SqlCommand objCommand = new SqlCommand(sql, conn);
                    conn.Open();
                        using (SqlDataReader objReader = objCommand.ExecuteReader(CommandBehavior.CloseConnection))
                        {
                            while (objReader.Read())
                            {
                                record record = new record();
                                record.Guid = Convert.ToString(objReader["guid"]);
                                record.Proj_name = Convert.ToString(objReader["proj_name"]);
                                records.Add(record);
                            }
                        }
                    conn.Close();
                    conn.Dispose();
                }
                return records;
            }


            /// <summary>
            ///根据 id 获取
            /// </summary>
            /// <param name="id"></param>
```

```csharp
/// <returns></returns>
public record GetrecordById(string guid)
{
    using (SqlConnection conn = new SqlConnection(connString))
    {
        record record = new record();
        string sql = "select * from T_record where guid='" + guid + "'";
        SqlCommand objCommand = new SqlCommand(sql, conn);
        conn.Open();
        using (SqlDataReader objReader = objCommand.ExecuteReader(CommandBehavior.CloseConnection))
        {
            if (objReader.Read())
            {
                record.Guid = Convert.ToString(objReader["guid"]);
                record.Proj_name = Convert.ToString(objReader["proj_name"]);
                record.Make_unit = Convert.ToString(objReader["make_unit"]);
                record.Time = Convert.ToDateTime(objReader["time"]);
                record.Result = Convert.ToString(objReader["result"]);
            }

        }
        return record;
    }
}

/// <summary>
///增加
/// </summary>
/// <param name="record"></param>
/// <returns>受影响的行数</returns>
public int Addrecord(record record)
{
    int number = 0;
    using (SqlConnection conn = new SqlConnection(connString))
    {
        string sql = "INSERT into T_record VALUES('" + record.Guid+"','" + record.Proj_name + "','" + record.Make_unit + "','" + record.Time + "','" + record.Result + "')";
        SqlCommand objCommand = new SqlCommand(sql, conn);
        conn.Open();

        number = objCommand.ExecuteNonQuery();
    }
```

```
            return number;
        }
    }
}

/**
 *实体类
 **/
using System;
using System.Collections.Generic;
using System.Linq;
using System.Text;

namespace Model
{
    public class record
    {
        private string guid;

        public string Guid
        {
            get { return guid; }
            set { guid = value; }
        }
        private string proj_name;

        public string Proj_name
        {
            get { return proj_name; }
            set { proj_name = value; }
        }

        private string make_unit;

        public string Make_unit
        {
            get { return make_unit; }
            set { make_unit = value; }
        }
        private DateTime time;

        public DateTime Time
        {
```

```csharp
            get { return time; }
            set { time = value; }
        }

        private string result;

        public string Result
        {
            get { return result; }
            set { result = value; }
        }

    }
}

/**
 *项目管理类
 **/
using System;
using System.Collections.Generic;
using System.Linq;
using System.Text;

using DAL;
using Model;

namespace BLL
{
    public class recordManager
    {

        recordServer server = new recordServer();

        public IList<record> GetAllrecord()
        {
            return server.GetAllrecord();
        }

        public record GetrecordById(string guid)
        {
            return server.GetrecordById(guid);
        }
```

```csharp
            public bool Addrecord(record record)
            {
                bool returns = false;
                try
                {
                    if (server.Addrecord(record) > 0)
                    {
                        returns = true;
                    }
                }
                catch (Exception)
                {
                    return returns;
                }
                return returns;
            }
        }
    }
```

项目十一

```csharp
/**
 *勘测定界资料管理类
 **/
using System;
using System.Collections.Generic;
using System.Linq;
using System.Text;
using Model;
using DAL;

namespace BLL
{
    public class AeraManager
    {
        #region Private Members
        AeraServer aeraServer = new AeraServer();
        #endregion

        public IList<Aera> GetAllInfo()
        {
            try
```

```csharp
            {
                return aeraServer.GetAllInfo();
            }
            catch (Exception ex)
            {
                throw new Exception(ex.ToString());
            }
        }

        public int ModifyAera(Aera aera)
        {
            int num;
            return num = aeraServer.ModifyAera(aera);
        }
    }
}
```

/**
*数据库操作类
**/

```csharp
using System;
using System.Collections.Generic;
using System.Linq;
using System.Text;
using Model;
using System.Configuration;
using System.Data.SqlClient;
using System.Data;



namespace DAL
{
    public class AeraServer
    {
        #region Private Members
        //从配置文件中读取数据库连接字符串
        private readonly string connString = ConfigurationManager.ConnectionStrings["ConnectionString"].ToString();
        private readonly string dboOwner = ConfigurationManager.ConnectionStrings["DataBaseOwner"].ToString();
        #endregion
```

```csharp
///<summary>
///返回所有信息集合
///</summary>
///<returns>信息集合</returns>
public IList<Aera> GetAllInfo()
{
    IList<Aera> aera = new List<Aera>();
    using (SqlConnection conn = new SqlConnection(connString))
    {
        string sql="select * from T_qualification";
        SqlCommand objCommand = new SqlCommand(sql, conn);
        conn.Open();
        using (SqlDataReader objReader = objCommand.ExecuteReader(CommandBehavior.CloseConnection))
        {
            while (objReader.Read())
            {
                Aera aera1 = new Aera();
                aera1.Ca_Guid= Convert.ToInt32(objReader["CA_GUID"]);
                aera1.Sb_Name = Convert.ToString(objReader["SB_NAME"]);
                aera1.Address = Convert.ToString(objReader["ADDRESS"]);
                aera1.Unit_Name = Convert.ToString(objReader["UNIT_NAME"]);
                aera1.Theowner= Convert.ToString(objReader["THEOWNER"]);
                aera.Add(aera1);
            }
        }
        conn.Close();
        conn.Dispose();
    }
    return aera;
}
///<summary>
///修改信息
///</summary>
///<param name="staff"></param>
///<returns></returns>

public int ModifyAera(Aera aera)
{
    int num;
    using (SqlConnection conn = new SqlConnection(connString))
    {
        string sql = string.Format("update T_qualification set SB_NAME='{0}',ADDRESS='{1}
```

',UNIT_NAME='{2}',THEOWNER='{3}' where CA_GUID={4} ", aera.Sb_Name, aera.Address, aera.Unit_Name, aera.Theowner, aera.Ca_Guid);
                SqlCommand objCommand = new SqlCommand(sql, conn);

                //objCommand.Parameters.Add("@StaffID", SqlDbType.Int).Value = staff.Staffid;
                //objCommand.Parameters.Add("@StaffLoginID", SqlDbType.VarChar, 10).Value = staff.Staffloginid;
                //objCommand.Parameters.Add("@StaffLoginPwd", SqlDbType.VarChar, 10).Value = staff.Staffloginpwd;
                conn.Open();
                num = objCommand.ExecuteNonQuery();
                conn.Close();
                conn.Dispose();
            }
            return num;
        }
    }
}

/**
 * 实体类
 **/
using System;
using System.Collections.Generic;
using System.Linq;
using System.Text;

namespace Model
{
    public class Aera
    {
        private int ca_Guid;

        public int Ca_Guid
        {
            get { return ca_Guid; }
            set { ca_Guid = value; }
        }
        private string sb_Name;

        public string Sb_Name
        {

```csharp
            get { return sb_Name; }
            set { sb_Name = value; }
        }
        private string address;
        public string Address
        {
            get { return address; }
            set { address = value; }
        }
        private string unit_Name;

        public string Unit_Name
        {
            get { return unit_Name; }
            set { unit_Name = value; }
        }
        private string theowner;

        public string Theowner
        {
            get { return theowner; }
            set { theowner = value; }
        }

    }
}


/**
*查询并修改所有勘测定界资料信息界面类
**/
using System;
using System.Collections.Generic;
using System.ComponentModel;
using System.Data;
using System.Drawing;
using System.Linq;
using System.Text;
using System.Windows.Forms;
using Model;
using BLL;


namespace 技能试题2
```

```csharp
{
    public partial class SelectOrModifyForm : Form
    {
        #region Private Members
        AeraManager aeraManager = new AeraManager();
        #endregion
        public SelectOrModifyForm()
        {
            InitializeComponent();
        }

        private void SelectForm_Load(object sender, EventArgs e)
        {
            datasource();
        }

        private void btnExit_Click(object sender, EventArgs e)
        {
            Application.Exit();
        }


        private void tsmiModify_Click(object sender, EventArgs e)
        {
            dgvInfo.ReadOnly = false;


        }
        /// <summary>
        ///为datagridView绑定数据源
        /// </summary>
        private void datasource()
        {
            dgvInfo.DataSource = null;
            dgvInfo.Rows.Clear();
            dgvInfo.DataSource = aeraManager.GetAllInfo();
            dgvInfo.Rows[0].Selected = false;
            dgvInfo.Columns[0].HeaderText = "分类编号";
            dgvInfo.Columns[1].HeaderText = "勘测定界资料名称";
            dgvInfo.Columns[2].HeaderText = "地址";
            dgvInfo.Columns[3].HeaderText = "单位名称";
            dgvInfo.Columns[4].HeaderText = "类别";
        }
```

```csharp
private void 保存ToolStripMenuItem_Click_1(object sender, EventArgs e)
{
    Aera aera = new Aera();
    int i = this.dgvInfo.SelectedRows[0].Index;
    aera.Ca_Guid = Convert.ToInt32(dgvInfo.Rows[i].Cells[0].Value.ToString().Trim());
    aera.Sb_Name = Convert.ToString(dgvInfo.Rows[i].Cells[1].Value.ToString().Trim());
    aera.Address = Convert.ToString(dgvInfo.Rows[i].Cells[2].Value.ToString().Trim());
    aera.Unit_Name = Convert.ToString(dgvInfo.Rows[i].Cells[3].Value.ToString().Trim());
    aera.Theowner = Convert.ToString(dgvInfo.Rows[i].Cells[4].Value.ToString().Trim());
    if (aeraManager.ModifyAera(aera) > 0)
    {
        MessageBox.Show("修改成功!", "提示信息", MessageBoxButtons.OK, MessageBoxIcon.Information);
    }
    else
        MessageBox.Show("修改失败!", "提示信息", MessageBoxButtons.OK, MessageBoxIcon.Information);
}

private void 取消ToolStripMenuItem_Click(object sender, EventArgs e)
{
    datasource();
    dgvInfo.ReadOnly = true;
}
}
}

/**
*启动类
**/
using System;
using System.Collections.Generic;
using System.Linq;
using System.Windows.Forms;

namespace 技能试题2
{
    static class Program
    {
        /// <summary>
        ///应用程序的主入口点。
        /// </summary>
```

```csharp
[STAThread]
static void Main()
{
    Application.EnableVisualStyles();
    Application.SetCompatibleTextRenderingDefault(false);
    Application.Run(new SelectOrModifyForm());
}
```

项目十二

```csharp
/**
 *数据库操作公共类
 **/
using System;
using System.Collections.Generic;
using System.Linq;
using System.Text;
using System.Data;
using System.Data.SqlClient;

namespace StudentManager
{
    public class DBHelper
    {
        public string str = @"Data Source=localhost\SA;Initial Catalog=StudentDB;Integrated Security=True";
        /// <summary>
        ///返回单个值
        /// </summary>
        /// <param name="sql"></param>
        /// <returns></returns>
        public object GetScalar(string sql)
        {
            using(SqlConnection con = new SqlConnection(str))
            using (SqlCommand com = new SqlCommand(sql, con))
            {
                con.Open();
                return com.ExecuteScalar();
            }
        }
```

```csharp
/// <summary>
///填充数据集
/// </summary>
public SqlDataAdapter da;
public void FillData(string sql,DataSet ds)
{
    SqlConnection con = new SqlConnection(str);
    da = new SqlDataAdapter(sql,con);
    da.Fill(ds);
}
}

/**
*登录类
**/
using System;
using System.Collections.Generic;
using System.ComponentModel;
using System.Data;
using System.Drawing;
using System.Linq;
using System.Text;
using System.Windows.Forms;

namespace StudentManager
{
    public partial class frmLogin : Form
    {
        public frmLogin()
        {
            InitializeComponent();
        }

        private void button2_Click(object sender, EventArgs e)
        {
            Application.Exit();
        }

        //登录
        private void button1_Click(object sender, EventArgs e)
        {
            string user = txtUser.Text.Trim();
```

```csharp
            string pwd = txtPwd.Text;
            if (user.Length == 0)
            {
                MessageBox.Show("请填写用户名","系统提示",MessageBoxButtons.OK,MessageBoxIcon.Error);
                txtUser.Focus();
                return;
            }
            if (pwd.Length == 0)
            {
                MessageBox.Show("请填写密码","系统提示",MessageBoxButtons.OK,MessageBoxIcon.Error);
                txtPwd.Focus();
                return;
            }
            DBHelper db = new DBHelper();
            string sql = string.Format("select user_password from T_user where User_name = '{0}'", user);
            try
            {
                string realPwd = Convert.ToString(db.GetScalar(sql));
                if (realPwd == pwd)
                {
                    frmStudentInfo frm = new frmStudentInfo();
                    frm.Show();
                    this.Hide();
                }
                else
                {
                    MessageBox.Show("用户名或密码有误","系统提示",MessageBoxButtons.OK,MessageBoxIcon.Error);
                    txtUser.Clear();
                    txtPwd.Clear();
                    txtUser.Focus();
                    return;
                }
            }
            catch (Exception ex)
            {
                MessageBox.Show("未知错误","系统提示",MessageBoxButtons.OK,MessageBoxIcon.Error);
            }
```

```
            }
        }
}

/**
 * 查询学生信息类
 * */
using System;
using System.Collections.Generic;
using System.ComponentModel;
using System.Data;
using System.Drawing;
using System.Linq;
using System.Text;
using System.Windows.Forms;

namespace StudentManager
{
    public partial class frmStudentInfo : Form
    {
        DataSet ds = new DataSet();
        DBHelper db = new DBHelper();
        public frmStudentInfo()
        {
            InitializeComponent();
        }

        //窗体加载时显示所有学生信息
        private void frmStudentInfo_Load(object sender, EventArgs e)
        {
            string sql = "select * from T_student_information";
            ShowStudent(sql);
        }

        //显示学生信息的方法
        public void ShowStudent(string sql)
        {
            //先清空原有记录
            if (ds.Tables.Count != 0)
            {
                ds.Tables[0].Clear();
            }
            db.FillData(sql, ds);
```

```csharp
            dgvStudent.DataSource = ds.Tables[0];

        }


        //取消
        private void button2_Click(object sender, EventArgs e)
        {
            Application.Exit();
        }


        //按条件查询
        private void btnOK_Click(object sender, EventArgs e)
        {
            string sql = "select * from T_student_information";
            string con = txtData.Text.Trim();
            if (con.Length == 0)
            {
                MessageBox.Show("请填写查询的条件", "系统提示", MessageBoxButtons.OK, MessageBoxIcon.Error);
                txtData.Focus();
                return;
            }
            if (rdoId.Checked == true) //按学号
            {
                sql += string.Format(" where student_id = '{0}'", con);
            }
            else if (rdoName.Checked == true)//按姓名
            {
                sql += string.Format(" where student_name = '{0}'", con);
            }
            else//按班级
            {
                sql += string.Format(" where class_id = '{0}'", con);
            }
            ShowStudent(sql);
            if (ds.Tables[0].Rows.Count == 0)
            {
                MessageBox.Show("无此记录", "系统提示", MessageBoxButtons.OK, MessageBoxIcon.Stop);
                return;
            }
        }
}
```

}

/**
 * 启动类
 * */
using System;
using System. Collections. Generic;
using System. Linq;
using System. Windows. Forms;

namespace StudentManager
{
    static class Program
    {
        /// <summary>
        ///应用程序的主入口点。
        /// </summary>
        [STAThread]
        static void Main()
        {
            Application. EnableVisualStyles();
            Application. SetCompatibleTextRenderingDefault(false);
            Application. Run(new frmLogin());
        }
    }
}

附录二　Java 方向部分参考答案

项目十三

代码如下：
数据库代码：
—— MySQL

```sql
DROP TABLE IF EXISTS `t_student_information`;
CREATE TABLE `t_student_information` (
  `Student_id` varchar(32) NOT NULL,
  `Student_name` varchar(64) default NULL,
  `Sex` varchar(64) default NULL,
  `Birthday` date default NULL,
  `Class_id` varchar(16) default NULL,
  `Telepone` varchar(32) default NULL,
  `Entry_date` date default NULL,
  `Address` varchar(50) default NULL,
  `Memo` varchar(50) default NULL,
  PRIMARY KEY (`Student_id`)
)

DROP TABLE IF EXISTS `t_user`;
CREATE TABLE `t_user` (
  `User_id` varchar(12) NOT NULL,
  `User_name` varchar(50) default NULL,
  `User_password` varchar(12) default NULL,
  PRIMARY KEY (`User_id`)
)
```

程序代码：

```java
package com.software;
import java.awt.*;
import java.awt.event.*;
import javax.swing.*;
import java.sql.*;
import java.awt.Frame;
```

```java
import java.awt.Rectangle;
import java.io.FileInputStream;
import sun.audio.*;
import java.io.*;
/**
 *定义修改密码弹出对话框窗体类
 **/
class xiugaimima extends JDialog {
    Statement ps;
    Statement slt;
    ResultSet rs;
    Connection con;
    String url;
    JPanel panel1 = new JPanel();
    JLabel jLabel1 = new JLabel();
    JTextField userF = new JTextField();
    JLabel jLabel2 = new JLabel();
    JButton cancel = new JButton();
    JButton sure = new JButton();
    JPasswordField pwd = new JPasswordField();
    JLabel jLabel3 = new JLabel();
    JPasswordField pwd1 = new JPasswordField();
    private javax.swing.JFileChooser jFileChooser1;
    public xiugaimima(Frame frame, String title, boolean modal) {
        super(frame, title, true);
        try {
            jbInit();
            pack();
        } catch (Exception ex) {
            ex.printStackTrace();
        }
    }
    public xiugaimima() {
        this(null, "修改密码", true);
    }
    /**
     * 初始化对话框窗体
     **/
    private void jbInit() throws Exception {
        panel1.setLayout(null);
        this.setModal(true);
        this.getContentPane().setLayout(null);
        panel1.setBounds(new Rectangle(-5, 0, 400, 300));
```

```
        jLabel1.setFont(new java.awt.Font("Dialog",0,15));
        jLabel1.setText("输入用户名");
        jLabel1.setBounds(new Rectangle(31,23,89,36));
        userF.setText("");
        userF.setBounds(new Rectangle(123,21,124,36));
        jLabel2.setFont(new java.awt.Font("Dialog",0,15));
        jLabel2.setText("输入新密码");
        jLabel2.setBounds(new Rectangle(30,69,78,38));
        cancel.setBounds(new Rectangle(148,169,89,35));
        cancel.setFont(new java.awt.Font("Dialog",0,15));
        cancel.setText("重填");
        cancel.addActionListener(new xiugaimima_cancel_actionAdapter(this));
        sure.setBounds(new Rectangle(46,167,88,37));
        sure.setFont(new java.awt.Font("Dialog",0,15));
        sure.setText("确定");
        sure.addActionListener(new xiugaimima_sure_actionAdapter(this));
        pwd.setText("");
        pwd.setBounds(new Rectangle(119,74,128,33));
        jLabel3.setBounds(new Rectangle(29,113,78,38));
        jLabel3.setText("再次输入");
        jLabel3.setFont(new java.awt.Font("Dialog",0,15));
        pwd1.setBounds(new Rectangle(119,116,128,33));
        pwd1.setText("");
        this.getContentPane().add(panel1,null);
        panel1.add(userF,null);
        panel1.add(jLabel1,null);
        panel1.add(jLabel2,null);
        panel1.add(pwd,null);
        panel1.add(pwd1,null);
        panel1.add(cancel,null);
        panel1.add(sure,null);
        panel1.add(jLabel3,null);
        this.setBounds(300,300,300,250);
        this.setVisible(true);
    }
    void cancel_actionPerformed(ActionEvent e){
        pwd.setText("");
        userF.setText("");
        pwd1.setText("");
    }
    void sure_actionPerformed(ActionEvent e){
        JOptionPane.showConfirmDialog(null,"确定修改","修改密码",
            JOptionPane.YES_NO_OPTION);
```

```java
            try {
                if (pwd.getText().trim().equals(pwd1.getText().trim())) {
                    try {
                        url = "jdbc:mysql://127.0.0.1:3306/studentDB";
                        Class.forName("com.mysql.jdbc.Driver");
                        Connection con = DriverManager.getConnection(url, "root",
                            "root");
                        ps = con.createStatement(ResultSet.TYPE_SCROLL_INSENSITIVE,
                            ResultSet.CONCUR_READ_ONLY);
                        slt = con.createStatement(ResultSet.TYPE_SCROLL_INSENSITIVE,
                            ResultSet.CONCUR_READ_ONLY);
                    } catch (Exception err) {
                        err.printStackTrace(System.out);
                    }
                    try {
                        if(ps.executeQuery("select * from T_user where User_name='"
                            + userF.getText().trim() + "'").next()==false)
                            {JOptionPane.showMessageDialog(null,"没有此用户");}
                        else
                            {ps.executeUpdate("Update T_user set User_password='"
                                + pwd.getText().trim() + "' where User_name='"
                                + userF.getText().trim() + "'");
                        JOptionPane.showMessageDialog(null,"修改成功");}
                        this.dispose();
                    } catch (SQLException sqle) {
                        String error = sqle.getMessage();
                        JOptionPane.showMessageDialog(null, error);
                        sqle.printStackTrace();
                    }
                } else {
                    JOptionPane.showMessageDialog(null,"两次密码不一致！请重新输入");
                }
            } catch (Exception ex) {
                ex.getMessage();
                String error = ex.getMessage();
                JOptionPane.showMessageDialog(null, error);
                ex.printStackTrace();
            }
        }
    void jButton1_actionPerformed(ActionEvent e) {
        if (JFileChooser.APPROVE_OPTION == jFileChooser1.showOpenDialog(this)) {
            String path = jFileChooser1.getSelectedFile().getPath();
            pwd.setText(path);
```

```java
            }
        }
    }
    class xiugaimima_cancel_actionAdapter implements java.awt.event.ActionListener {
        xiugaimima adaptee;
        xiugaimima_cancel_actionAdapter(xiugaimima adaptee) {
            this.adaptee = adaptee;
        }
        public void actionPerformed(ActionEvent e) {
            adaptee.cancel_actionPerformed(e);
        }
    }
    class xiugaimima_sure_actionAdapter implements java.awt.event.ActionListener {
        xiugaimima adaptee;
        xiugaimima_sure_actionAdapter(xiugaimima adaptee) {
            this.adaptee = adaptee;
        }
        public void actionPerformed(ActionEvent e) {
            adaptee.sure_actionPerformed(e);
        }
    }
    /**
     * 程序运行主窗体类
     */
    public class mainFrame extends JFrame {
        JPanel contentPane;
        String user1;
        int power;
        Statement ps;
        ResultSet rs;
        Connection con;
        String url;
        String username;
        JMenuBar jMenuBar1 = new JMenuBar();
        JMenu jMenuFile = new JMenu();
        JMenu jMenuHelp = new JMenu();
        JMenuItem jMenuHelpAbout = new JMenuItem();
        JMenuItem adduser = new JMenuItem();
        JMenu xjgl = new JMenu();
        JMenuItem cxxj = new JMenuItem();
        JMenuItem xgxj = new JMenuItem();
        JMenuItem tjxj = new JMenuItem();
```

```java
JMenu bjgl = new JMenu();
JMenuItem xgbj = new JMenuItem();
JMenuItem tjbj = new JMenuItem();
JMenu kcsz = new JMenu();
JMenuItem sznj = new JMenuItem();
JMenuItem xgkc = new JMenuItem();
JMenuItem tjkc = new JMenuItem();
JMenu cjgl = new JMenu();
JMenuItem tjcj = new JMenuItem();
ImageIcon icon = new ImageIcon("images" + File.separator + "a.jpg");
JLabel jLabel1 = new JLabel(icon, JLabel.CENTER);
JMenuItem xgcj = new JMenuItem();
JMenuItem cxcj = new JMenuItem();
JMenuItem exit = new JMenuItem();
JMenuItem jMenuItem1 = new JMenuItem();
JMenuItem llyh = new JMenuItem();
JMenuItem login = new JMenuItem();
JMenu ghbj = new JMenu();
JMenuItem jMenuItem2 = new JMenuItem();
JMenuItem jMenuItem3 = new JMenuItem();
JMenuItem jMenuItem4 = new JMenuItem();
JMenuItem jMenuItem5 = new JMenuItem();
FileInputStream fileau;// = new FileInputStream("lzlh.mid");
AudioStream as;
JMenu jMenu1 = new JMenu();
JMenuItem jMenuItem6 = new JMenuItem();
JMenuItem jMenuItem7 = new JMenuItem();
public mainFrame() {
    enableEvents(AWTEvent.WINDOW_EVENT_MASK);
    try {
        jbInit();
    } catch (Exception e) {
        e.printStackTrace();
    }
}
private void jbInit() throws Exception {
    contentPane = (JPanel) this.getContentPane();
    contentPane.setLayout(null);
    this.setResizable(false);
    this.setTitle("学生管理系统");
    jMenuFile.setFont(new java.awt.Font("Dialog", 0, 15));
    jMenuFile.setForeground(Color.black);
    jMenuFile.setText("
```

系统 ");
        jMenuHelp.setFont(new java.awt.Font("Dialog",0,15));
        jMenuHelp.setText(" 帮助
");
        jMenuHelpAbout.setFont(new java.awt.Font("Dialog",0,15));
        jMenuHelpAbout.setText("关于
");
        adduser.setText("添加用户");
        xjgl.setFont(new java.awt.Font("Dialog",0,15));
        xjgl.setText("
学籍管理
");
        bjgl.setFont(new java.awt.Font("Dialog",0,15));
        bjgl.setText("
班级管理
");
        kcsz.setFont(new java.awt.Font("Dialog",0,15));
        kcsz.setText("
课程设置
");
        cjgl.setFont(new java.awt.Font("Dialog",0,15));
        cjgl.setText("成绩管理");
        exit.setFont(new java.awt.Font("Dialog",0,15));
        exit.setText("退出");
        exit.addActionListener(new mainFrame_exit_actionAdapter(this));
        exit.addMouseListener(new mainFrame_exit_mouseAdapter(this));
        jMenuItem1.setFont(new java.awt.Font("Dialog",0,15));
        jMenuItem1.setText("修改密码");
        jMenuItem1.addActionListener(new mainFrame_jMenuItem1_actionAdapter(
            this));
        llyh.setFont(new java.awt.Font("Dialog",0,15));
        llyh.setText("浏览用户");
        login.setFont(new java.awt.Font("Dialog",0,15));
        login.setText("用户登录");
        ghbj.setFont(new java.awt.Font("Dialog",0,15));
        ghbj.setText("更换背景");
        jMenu1.addActionListener(new mainFrame_jMenu1_actionAdapter(this));
        jMenu1.setFont(new java.awt.Font("Dialog",0,15));
        jMenu1.setText("背景音乐");
        jMenuFile.add(login);
        jMenuFile.add(llyh);
        jMenuFile.add(jMenuItem1);
        jMenuFile.add(adduser);

```
            jMenuFile. add(exit);
            jMenuHelp. add(jMenuHelpAbout);
            jMenuBar1. add(jMenuFile);
            jMenuBar1. add(xjgl);
            jMenuBar1. add(bjgl);
            jMenuBar1. add(kcsz);
            jMenuBar1. add(cjgl);
            jMenuBar1. add(ghbj);
            jMenuBar1. add(jMenu1);
            jMenuBar1. add(jMenuHelp);
            xjgl. add(tjxj);
            xjgl. add(xgxj);
            xjgl. add(cxxj);
            bjgl. add(tjbj);
            bjgl. add(xgbj);
            kcsz. add(tjkc);
            kcsz. add(xgkc);
            kcsz. add(sznj);
            cjgl. add(tjcj);
            cjgl. add(xgcj);
            cjgl. add(cxcj);
            contentPane. add(jLabel1, null);
            ghbj. add(jMenuItem4);
            ghbj. add(jMenuItem3);
            ghbj. add(jMenuItem2);
            ghbj. add(jMenuItem5);
            jMenu1. add(jMenuItem7);
            jMenu1. add(jMenuItem6);
            this. setJMenuBar(jMenuBar1);
            xjgl. setEnabled(false);
            bjgl. setEnabled(false);
            kcsz. setEnabled(false);
            cjgl. setEnabled(false);
            jMenuFile. setEnabled(true);
            llyh. setEnabled(false);
            adduser. setEnabled(false);
            jMenuItem7. setEnabled(false);
            this. setBounds(100, 100, 800, 600);
            this. setVisible(true);
    }
       public static void main(String args[]) {
            mainFrame main = new mainFrame();
       }
```

```java
void xgmm_actionPerformed(ActionEvent e) {
    new xiugaimima();
}
void exit_actionPerformed(ActionEvent e) {
    System.exit(0);
}
void jMenuItem1_actionPerformed(ActionEvent e) {
    new xiugaimima();
}
void jMenuItem4_actionPerformed(ActionEvent e) {
    ImageIcon icon2 = new ImageIcon("images" + File.separator
            + "forest.jpg");
    jLabel1.setIcon(icon2);
    contentPane.add(jLabel1, null);
}
void jMenuItem2_actionPerformed(ActionEvent e) {
    ImageIcon icon2 = new ImageIcon("images" + File.separator + "jgs.jpg");
    jLabel1.setIcon(icon2);
    contentPane.add(jLabel1, null);
}
void jMenuItem5_actionPerformed(ActionEvent e) {
    ImageIcon icon2 = new ImageIcon("images" + File.separator
            + "shuijing.jpg");
    jLabel1.setIcon(icon2);
    contentPane.add(jLabel1, null);
}
void jMenuItem3_actionPerformed(ActionEvent e) {
    ImageIcon icon2 = new ImageIcon("images" + File.separator
            + "Autumn.jpg");
    jLabel1.setIcon(icon2);
    contentPane.add(jLabel1, null);
}
void jMenu1_actionPerformed(ActionEvent e) {
}
void jMenuItem7_actionPerformed(ActionEvent e) {
    jMenuItem7.setEnabled(false);
}
void jMenuItem6_actionPerformed(ActionEvent e) {
    AudioPlayer.player.stop(as);
    jMenuItem7.setEnabled(true);
}
}
```

```java
class mainFrame_jMenuHelpAbout_ActionAdapter implements ActionListener {
    mainFrame adaptee;
    mainFrame_jMenuHelpAbout_ActionAdapter(mainFrame adaptee) {
        this.adaptee = adaptee;
    }
    public void actionPerformed(ActionEvent e) {
    }
}

class mainFrame_adduser_actionAdapter implements java.awt.event.ActionListener {
    mainFrame adaptee;
    mainFrame_adduser_actionAdapter(mainFrame adaptee) {
        this.adaptee = adaptee;
    }
    public void actionPerformed(ActionEvent e) {
    }
}

class mainFrame_tjbj_actionAdapter implements java.awt.event.ActionListener {
    mainFrame adaptee;
    mainFrame_tjbj_actionAdapter(mainFrame adaptee) {
        this.adaptee = adaptee;
    }
    public void actionPerformed(ActionEvent e) {
    }
}

class mainFrame_exit_mouseAdapter extends java.awt.event.MouseAdapter {
    mainFrame adaptee;
    mainFrame_exit_mouseAdapter(mainFrame adaptee) {
        this.adaptee = adaptee;
    }
}

class mainFrame_xgbj_actionAdapter implements java.awt.event.ActionListener {
    mainFrame adaptee;
    mainFrame_xgbj_actionAdapter(mainFrame adaptee) {
        this.adaptee = adaptee;
    }
    public void actionPerformed(ActionEvent e) {
    }
```

}

class mainFrame_exit_actionAdapter implements java.awt.event.ActionListener {
    mainFrame adaptee;
    mainFrame_exit_actionAdapter(mainFrame adaptee) {
        this.adaptee = adaptee;
    }
    public void actionPerformed(ActionEvent e) {
        adaptee.exit_actionPerformed(e);
    }
}

class mainFrame_xjgl_actionAdapter implements java.awt.event.ActionListener {
    mainFrame adaptee;
    mainFrame_xjgl_actionAdapter(mainFrame adaptee) {
        this.adaptee = adaptee;
    }
    public void actionPerformed(ActionEvent e) {
    }
}

class mainFrame_tjxj_actionAdapter implements java.awt.event.ActionListener {
    mainFrame adaptee;
    mainFrame_tjxj_actionAdapter(mainFrame adaptee) {
        this.adaptee = adaptee;
    }
    public void actionPerformed(ActionEvent e) {
    }
}

class mainFrame_xgxj_actionAdapter implements java.awt.event.ActionListener {
    mainFrame adaptee;
    mainFrame_xgxj_actionAdapter(mainFrame adaptee) {
        this.adaptee = adaptee;
    }
    public void actionPerformed(ActionEvent e) {
    }
}

class mainFrame_cxxj_actionAdapter implements java.awt.event.ActionListener {
    mainFrame adaptee;
    mainFrame_cxxj_actionAdapter(mainFrame adaptee) {
        this.adaptee = adaptee;

```
        }
        public void actionPerformed(ActionEvent e) {
        }
}
class mainFrame_tjkc_actionAdapter implements java.awt.event.ActionListener {
    mainFrame adaptee;
    mainFrame_tjkc_actionAdapter(mainFrame adaptee) {
        this.adaptee = adaptee;
    }
    public void actionPerformed(ActionEvent e) {
    }
}
class mainFrame_llyh_actionAdapter implements java.awt.event.ActionListener {
    mainFrame adaptee;
    mainFrame_llyh_actionAdapter(mainFrame adaptee) {
        this.adaptee = adaptee;
    }

    public void actionPerformed(ActionEvent e) {

    }
}
class mainFrame_jMenuItem1_actionAdapter implements
        java.awt.event.ActionListener {
    mainFrame adaptee;

    mainFrame_jMenuItem1_actionAdapter(mainFrame adaptee) {
        this.adaptee = adaptee;
    }
    public void actionPerformed(ActionEvent e) {
        adaptee.jMenuItem1_actionPerformed(e);
    }
}

class mainFrame_xgkc_actionAdapter implements java.awt.event.ActionListener {
    mainFrame adaptee;
    mainFrame_xgkc_actionAdapter(mainFrame adaptee) {
        this.adaptee = adaptee;
    }
    public void actionPerformed(ActionEvent e) {
```

```java
    }
}

class mainFrame_sznj_actionAdapter implements java.awt.event.ActionListener {
    mainFrame adaptee;
    mainFrame_sznj_actionAdapter(mainFrame adaptee) {
        this.adaptee = adaptee;
    }
    public void actionPerformed(ActionEvent e) {

    }
}

class mainFrame_tjcj_actionAdapter implements java.awt.event.ActionListener {
    mainFrame adaptee;
    mainFrame_tjcj_actionAdapter(mainFrame adaptee) {
        this.adaptee = adaptee;
    }
    public void actionPerformed(ActionEvent e) {
    }
}

class mainFrame_xgcj_actionAdapter implements java.awt.event.ActionListener {
    mainFrame adaptee;
    mainFrame_xgcj_actionAdapter(mainFrame adaptee) {
        this.adaptee = adaptee;
    }
    public void actionPerformed(ActionEvent e) {
    }
}

class mainFrame_cxcj_actionAdapter implements java.awt.event.ActionListener {
    mainFrame adaptee;
    mainFrame_cxcj_actionAdapter(mainFrame adaptee) {
        this.adaptee = adaptee;
    }
    public void actionPerformed(ActionEvent e) {
    }
}

class mainFrame_login_actionAdapter implements java.awt.event.ActionListener {
    mainFrame adaptee;
    mainFrame_login_actionAdapter(mainFrame adaptee) {
```

```java
            this.adaptee = adaptee;
        }
        public void actionPerformed(ActionEvent e) {
        }
    }

    class mainFrame_jMenuItem4_actionAdapter implements
            java.awt.event.ActionListener {
        mainFrame adaptee;
        mainFrame_jMenuItem4_actionAdapter(mainFrame adaptee) {
            this.adaptee = adaptee;
        }
        public void actionPerformed(ActionEvent e) {
            adaptee.jMenuItem4_actionPerformed(e);
        }
    }

    class mainFrame_jMenuItem2_actionAdapter implements
            java.awt.event.ActionListener {
        mainFrame adaptee;
        mainFrame_jMenuItem2_actionAdapter(mainFrame adaptee) {
            this.adaptee = adaptee;
        }

        public void actionPerformed(ActionEvent e) {
            adaptee.jMenuItem2_actionPerformed(e);
        }
    }

    class mainFrame_jMenuItem5_actionAdapter implements
            java.awt.event.ActionListener {
        mainFrame adaptee;
        mainFrame_jMenuItem5_actionAdapter(mainFrame adaptee) {
            this.adaptee = adaptee;
        }
        public void actionPerformed(ActionEvent e) {
            adaptee.jMenuItem5_actionPerformed(e);
        }
    }

    class mainFrame_jMenuItem3_actionAdapter implements
            java.awt.event.ActionListener {
        mainFrame adaptee;
```

```java
    mainFrame_jMenuItem3_actionAdapter(mainFrame adaptee) {
        this.adaptee = adaptee;
    }
    public void actionPerformed(ActionEvent e) {
        adaptee.jMenuItem3_actionPerformed(e);
    }
}

class mainFrame_jMenu1_actionAdapter implements java.awt.event.ActionListener {
    mainFrame adaptee;
    mainFrame_jMenu1_actionAdapter(mainFrame adaptee) {
        this.adaptee = adaptee;
    }
    public void actionPerformed(ActionEvent e) {
        adaptee.jMenu1_actionPerformed(e);
    }
}

class mainFrame_jMenuItem7_actionAdapter implements
        java.awt.event.ActionListener {
    mainFrame adaptee;
    mainFrame_jMenuItem7_actionAdapter(mainFrame adaptee) {
        this.adaptee = adaptee;
    }
    public void actionPerformed(ActionEvent e) {
        adaptee.jMenuItem7_actionPerformed(e);
    }
}

class mainFrame_jMenuItem6_actionAdapter implements
        java.awt.event.ActionListener {
    mainFrame adaptee;
    mainFrame_jMenuItem6_actionAdapter(mainFrame adaptee) {
        this.adaptee = adaptee;
    }
    public void actionPerformed(ActionEvent e) {
        adaptee.jMenuItem6_actionPerformed(e);
    }
    public static void main(String args[]) {
        new mainFrame();
    }
}
```

项目十四

代码如下：
创建数据库代码

```sql
CREATE DATABASE  HNIUEAMDB CHARACTER SET gbk;
USE HNIUEAMDB;
CREATE TABLE t_supplier_information (
    Supplier_id VARCHAR (10) NOT NULL PRIMARY KEY,  —— 供应商编号（主键）
    Supplier_name VARCHAR (50) NOT NULL,            —— 供应商名称
    Supplier_people VARCHAR (8) NOT NULL,           —— 供应商联系人
    Supplier_address VARCHAR (50) NULL,             —— 供应商地址
    Supplier_phone VARCHAR (11) NULL,               —— 供应商电话
    Supplier_code VARCHAR (6) NULL                  —— 供应商邮编
);
CREATE TABLE t_order (
    Order_id VARCHAR (10) NOT NULL PRIMARY KEY,     —— 订单编号
    Supplier_id VARCHAR (10) NOT NULL,              —— 供应商ID
    Order_date datetime NOT NULL,                   —— 订货日期
    Order_status INT NOT NULL                       —— 订单状态
);
Insert into
t_supplier_information(Supplier_id,Supplier_name,Supplier_people,Supplier_address,Supplier_phone,Supplier_code)
    select 'BJ1002','清华大学出版社','郭政强','北京清华大学','15123467890','101023'  UNION
    select 'CD1003','科技出版社','蒋军','成都电子科技大学','15874679856','290897'  UNION
    select 'CS1001','湖南大学出版社','李伟','长沙市岳麓区湖南大学','13789654673','410230';
insert into t_order(Order_id,Supplier_id,Order_date,Order_status)
    select 'DD1001201','CS1001','2010-03-02',1 UNION
    select 'DD1001202','BJ1002','2011-04-04',0 UNION
    select 'DD1001203','CD1003','2008-08-08',1;
```

DBUtil.java 代码

```java
package com.software.dao;
import java.sql.Connection;
import java.sql.DriverManager;
import java.sql.SQLException;
/**
 * 类名：DBUtil <br/>
 * 功能：自定义数据库工具类,封装通用数据库操作 <br/>
 * 创建时间：2016-5-31 下午 4:51:22 <br/>
```

* @author Administrator
 * @version
 * @since JDK 1.6
 */
public class DBUtil {
    /**
     * getConnection：连接数据库返回，数据库连接对象。
     * @author Administrator
     * @return 数据库连接对象
     * @throws ClassNotFoundException
     * @throws SQLException
     * @since JDK 1.6
     */
    public static Connection getConnection() throws ClassNotFoundException,SQLException{
        String url="jdbc:mysql://127.0.0.1:8306/HNIUEAMDB";
        String user="root";
        String pwd="1234";
        Class.forName("com.mysql.jdbc.Driver");
        Connection conn=DriverManager.getConnection(url, user, pwd);
        return conn;
    }
}
```

MyTableModel.java 代码

```java
package com.software.dao;
import java.util.Vector;
import javax.swing.table.DefaultTableModel;
/**
 * 类名：MyTableModel <br/>
 * 功能：自定义 JTable 表格数据模型代码。<br/>
 * 创建时间：2016-5-31 下午 5:03:10 <br/>
 * @author Administrator
 * @version
 * @since JDK 1.6
 */
public class MyTableModel extends DefaultTableModel {
    /**
     * 创建一个新的实例 MyTableModel.
     * @param data 表格数据
     * @param columnNames 显示表格的列名
     */

```java
        public MyTableModel(Vector data, Vector columnNames) {
            super(data, columnNames);
        }
        /**
         * 控制表格单元格是否可编辑.
         * @param r 行号
         * @param c 列号
         */
        public boolean isCellEditable(int r, int c) {
            return false;
        }
        /**
         * 获得单元格列类型.
         * @param c 列号
         * @return 列类型
         */
        public Class getColumnClass(int c) {
            return getValueAt(0, c).getClass();
        }
    }
}
```

OrderDao.java 代码

```java
package com.software.dao;
import java.sql.Connection;
import java.sql.PreparedStatement;
import java.sql.ResultSet;
import java.sql.SQLException;
import java.sql.Statement;
import java.util.Vector;
import javax.swing.table.DefaultTableModel;
import javax.swing.table.TableModel;
import static com.software.dao.DBUtil.*;
/**
 * 类名：OrderDao <br/>
 * 功能：封装对订单表数据库操作. <br/>
 * 创建时间：2016-5-31 下午 5:18:48 <br/>
 * @author Administrator
 * @version
 * @since JDK 1.6
 */
public class OrderDao {
    /**
     * getColumnNames:获取中文显示的表格列标题. <br/>
     * @author
```

Administrator
* @return 返回中文显示的表格标题
* @since JDK 1.6
*/
public static Vector<String> getColumnNames(){
    Vector<String> columnNames=new Vector<String>();
    columnNames.add(0,"订单 ID");
    columnNames.add(1,"供应商名称");
    columnNames.add(2,"订单日期");
    columnNames.add(3,"订单状态");
    return columnNames;
}
/**
* getEmpty:获取一个带中文列标题,但数据为空表格数据模型.
* @author
Administrator
* @return 返回一个带中文列标题,但数据为空表格数据模型
* @since JDK 1.6
*/
public static TableModel getEmpty(){
    Vector data=new Vector();
    DefaultTableModel dmt=new DefaultTableModel(data, getColumnNames());
    return dmt;
}
/**
* queryOrders:查询订单数据.
* @author
Administrator
* @param orderId
* @return 返回查询订单数据
* @since JDK 1.6
*/
public Vector queryOrders(String orderId){
    Vector data=new Vector();
    Connection conn=null;
    Statement stm=null;
    ResultSet rs=null;
    String sql="select order_id,supplier_id,"+
        "(select supplier_name from t_supplier_information a where a.Supplier_id=b.Supplier_id) supplier_name,order_date,order_status"+
        " from t_order b ";
    try {
        if(!(orderId==null||orderId.equals(""))){

```java
                sql+=" where order_id='"+orderId+"'";
            }
            conn=getConnection();
            stm=conn.createStatement();
            rs=stm.executeQuery(sql);
            while(rs.next()){
                Vector row=new Vector();
                row.add(rs.getString(1));
                row.add(rs.getString(3));
                row.add(rs.getString(4).substring(0,10));
                if(rs.getInt(5)==1) row.add(true);
                else row.add(false);
                data.add(row);
            }
                } catch (ClassNotFoundException e) {
            e.printStackTrace();
        } catch (SQLException e) {
            System.out.println(sql);
            e.printStackTrace();
        }finally{
            try {
                if(rs!=null) rs.close();
                if(stm!=null)stm.close();
                if(conn!=null) conn.close();
            } catch (SQLException e) {
                    e.printStackTrace();
            }
        }
        return data;
    }
}
```

OrderDao.java 代码

```java
package com.software.dao;
import static com.software.dao.DBUtil.*;
import java.sql.Connection;
import java.sql.PreparedStatement;
import java.sql.SQLException;
import com.software.entity.SupplierBean;
/**
 * 类名：SupplierDao <br/>
 * 功能：封装对供应商表的数据库操作。<br/>
 * 创建时间：2016-5-31 下午 5:30:39 <br/>
```

```java
 * @author Administrator
 * @version
 * @since JDK 1.6
 */
public class SupplierDao {
    /**
     * addSupplier:往供应商表中添加一条记录.<br/>
     * @author Administrator
     * @param supplier
     * @return 是否添加成功
     * @since JDK 1.6
     */
    public boolean addSupplier(SupplierBean supplier){
        boolean isSuccess=false;
        Connection conn=null;
        PreparedStatement ps=null;
        try {
            String sql="insert into t_supplier_information values(?,?,?,?,?,?)";
            conn=getConnection();
            ps=conn.prepareStatement(sql);
            ps.setString(1, supplier.getSupplierId());
            ps.setString(2, supplier.getSupplierName());
            ps.setString(3, supplier.getSupplierPeople());
            ps.setString(4, supplier.getSupplierAddress());
            ps.setString(5, supplier.getSupplierPhone());
            ps.setString(6, supplier.getSupplierCode());
            ps.executeUpdate();
            isSuccess=true;
        } catch (ClassNotFoundException e) {
            e.printStackTrace();
        } catch (SQLException e) {
            e.printStackTrace();
        }finally{
            try {
                if(ps!=null) ps.close();
                if(conn!=null) conn.close();
            } catch (SQLException e) {
                e.printStackTrace();
            }
        }
        return isSuccess;
    }
```

}

SupplierBean.java 代码

```java
package com.software.entity;
/**
 * 类名：SupplierBean <br/>
 * 功能：对应供应商表的实体类. <br/>
 * 创建时间：2016-5-31 下午 5:35:00 <br/>
 * @author Administrator
 * @version
 * @since JDK 1.6
 */
public class SupplierBean {
    private String supplierId;
    private String supplierName;
    private String supplierPeople;
    private String supplierAddress;
    private String supplierPhone;
    private String supplierCode;
    public String getSupplierId() {
        return supplierId;
    }
    public void setSupplierId(String supplierId) {
        this.supplierId = supplierId;
    }
    public String getSupplierName() {
        return supplierName;
    }
    public void setSupplierName(String supplierName) {
        this.supplierName = supplierName;
    }
    public String getSupplierPeople() {
        return supplierPeople;
    }
    public void setSupplierPeople(String supplierPeople) {
        this.supplierPeople = supplierPeople;
    }
    public String getSupplierAddress() {
        return supplierAddress;
    }
    public void setSupplierAddress(String supplierAddress) {
        this.supplierAddress = supplierAddress;
    }
    public String getSupplierPhone() {
```

```java
        return supplierPhone;
    }
    public void setSupplierPhone(String supplierPhone) {
        this.supplierPhone = supplierPhone;
    }
    public String getSupplierCode() {
        return supplierCode;
    }
    public void setSupplierCode(String supplierCode) {
        this.supplierCode = supplierCode;
    }
}
```

AddPanel.java 代码

```java
package com.software.ui;
import java.awt.event.ActionEvent;
import java.awt.event.ActionListener;
import javax.swing.JButton;
import javax.swing.JLabel;
import javax.swing.JOptionPane;
import javax.swing.JPanel;
import javax.swing.JTextField;
import javax.swing.SwingConstants;
import com.software.dao.SupplierDao;
import com.software.entity.SupplierBean;
/**
 * 类名：AddPanel <br/>
 * 功能：增加供应商信息界面代码。<br/>
 * 创建时间：2016-5-31 下午 5:38:42 <br/>
 * @author Administrator
 * @version
 * @since JDK 1.6
 */
public class AddPanel extends JPanel {
    JLabel jlb_Id=new JLabel("供应商 ID:",SwingConstants.LEFT);
    JLabel jlb_Name=new JLabel("供应商名称:",SwingConstants.LEFT);
    JLabel jlb_People=new JLabel("联系人:",SwingConstants.LEFT);
    JLabel jlb_Address=new JLabel("地址:",SwingConstants.LEFT);
    JLabel jlb_Phone=new JLabel("电话:",SwingConstants.LEFT);
    JLabel jlb_Code=new JLabel("邮编:",SwingConstants.LEFT);
        JTextField jtf_Id=new JTextField(30);
    JTextField jtf_Name=new JTextField(30);
    JTextField jtf_People=new JTextField(30);
```

```java
JTextField jtf_Address=new JTextField(30);
JTextField jtf_Phone=new JTextField(30);
JTextField jtf_Code=new JTextField(30);
JButton jb_Add=new JButton("添加");
/**
 * 创建一个新的实例 AddPanel 对界面控件进行布局.
 */
public AddPanel(){
    this.setLayout(null);
    this.setSize(300, 300);
    add(jlb_Id); add(jtf_Id);
    jlb_Id.setBounds(20,20, 80,30);
    jtf_Id.setBounds(100, 20, 200, 30);
    add(jlb_Name); add(jtf_Name);
    jlb_Name.setBounds(20, 60, 80, 30);
    jtf_Name.setBounds(100, 60, 200,30);
    add(jlb_People);add(jtf_People);
    jlb_People.setBounds(20, 100, 80, 30);
    jtf_People.setBounds(100,100,200,30);
    add(jlb_Address);add(jtf_Address);
    jlb_Address.setBounds(20, 140, 80, 30);
    jtf_Address.setBounds(100, 140, 200, 30);
    add(jlb_Phone);add(jtf_Phone);
    jlb_Phone.setBounds(20,180, 80,30);
    jtf_Phone.setBounds(100, 180,200, 30);
    add(jlb_Code);add(jtf_Code);
    jlb_Code.setBounds(20, 220, 80, 30);
    jtf_Code.setBounds(100,220,200, 30);
    add(jb_Add);
    jb_Add.setBounds(100, 260, 200, 30);
    AddHandler addHandler=new AddHandler();
    jb_Add.addActionListener(addHandler);
}
/**
 * 类名：AddHandler <br/>
 * 功能：添加按钮的事件处理类. <br/>
 * 创建时间：2016-5-31 下午5:41:19 <br/>
 * @author Administrator
 * @version AddPanel
 * @since JDK 1.6
 */
class AddHandler implements ActionListener{
    public void actionPerformed(ActionEvent e) {
```

```java
            SupplierBean supplier=new SupplierBean();
            supplier.setSupplierId(jtf_Id.getText());
            supplier.setSupplierName(jtf_Name.getText());
            supplier.setSupplierPeople(jtf_People.getText());
            supplier.setSupplierAddress(jtf_Address.getText());
            supplier.setSupplierPhone(jtf_Phone.getText());
            supplier.setSupplierCode(jtf_Code.getText());
            SupplierDao dao=new SupplierDao();
            if(dao.addSupplier(supplier)){
                JOptionPane.showMessageDialog(null,"添加供应商信息成功");
                jtf_Id.setText("");
                jtf_Name.setText("");
                jtf_People.setText("");
                jtf_Address.setText("");
                jtf_Phone.setText("");
                jtf_Code.setText("");
            }else{
                JOptionPane.showMessageDialog(null,"添加供应商信息失败");
            }
        }
    }
}
```

QueryPanel.java 代码

```java
package com.software.ui;
import java.awt.BorderLayout;
import java.awt.event.ActionEvent;
import java.awt.event.ActionListener;
import java.util.Vector;
import javax.swing.BorderFactory;
import javax.swing.JButton;
import javax.swing.JLabel;
import javax.swing.JOptionPane;
import javax.swing.JPanel;
import javax.swing.JScrollPane;
import javax.swing.JTable;
import javax.swing.JTextField;
import javax.swing.SwingConstants;
import com.software.dao.MyTableModel;
import com.software.dao.OrderDao;
import static com.software.dao.OrderDao.*;
/**
 * 类名：QueryPanel <br/>
```

```java
 * 功能：查询订单信息的界面代码.<br/>
 * 创建时间：2016-5-31 下午5:45:03 <br/>
 * @author Administrator
 * @version
 * @since JDK 1.6
 */
public class QueryPanel extends JPanel {
    JPanel jp_condition=new JPanel();
    JLabel jlb_Id=new JLabel("订单编号：",SwingConstants.RIGHT);
    JTextField jtf_Id=new JTextField(30);
    JButton jb_query=new JButton("查询");
    JTable jtable_query=new JTable();
    JScrollPane jScrollPane=new JScrollPane();
    /**
     * 创建一个新的实例 QueryPanel 对界面控件进行布局.
     */
    public QueryPanel(){
        setSize(400,300);
        setLayout(new BorderLayout());
        jtable_query.setModel(getEmpty());
        jp_condition.add(jlb_Id);
        jp_condition.add(jtf_Id);
        jp_condition.add(jb_query);
        jp_condition.setBorder(BorderFactory.createTitledBorder("===查询条件==="));
        jScrollPane.setBorder(BorderFactory.createTitledBorder("===订单信息==="));
        jScrollPane.setViewportView(jtable_query);
        add(jp_condition,BorderLayout.NORTH);
        add(jScrollPane);
        jb_query.addActionListener(new QueryHandler());
    }
    /**
     * 类名：QueryHandler <br/>
     * 功能：查询按钮的事件处理类. <br/>
     * 创建时间：2016-5-31 下午5:46:25 <br/>
     * @author Administrator
     * @version QueryPanel
     * @since JDK 1.6
     */
    class QueryHandler implements ActionListener{
        public void actionPerformed(ActionEvent e) {
            OrderDao dao=new OrderDao();
            Vector data=dao.queryOrders(jtf_Id.getText());
```

```
            if(data.size()<1){
                JOptionPane.showMessageDialog(null,"没有查询到数据,请更改查询条件!");
            }else{
                MyTableModel mt=new MyTableModel(data,getColumnNames());
                jtable_query.setModel(mt);
            }
        }
    }
}
```

MainFrame.java 代码

```
package com.software.ui;
import java.awt.Container;
import javax.swing.JFrame;
import javax.swing.JTabbedPane;
/**
 * 类名：MainFrame <br/>
 * 功能：主窗体类代码用于加载面板. <br/>
 * 创建时间：2016-5-31 下午 5:50:08 <br/>           *
 * @author Administrator
 * @version
 * @since JDK 1.6
 */
public class MainFrame extends JFrame {
    /**
     * 创建一个新的实例 MainFrame,加载面板进行界面布局.
     */
    public MainFrame(){
        AddPanel addPanel=new AddPanel();
        QueryPanel queryPanel=new QueryPanel();
        Container content=this.getContentPane();
            JTabbedPane tabbedPane=new JTabbedPane();
        tabbedPane.addTab("添加供应商信息",addPanel);
        tabbedPane.addTab("订单查询",queryPanel);
        content.add(tabbedPane);
        this.setTitle("订单管理");
        this.setSize(550,400);
        this.setVisible(true);
        this.setDefaultCloseOperation(JFrame.EXIT_ON_CLOSE);
    }
    /**
     * main:程序运行主函数. <br/>
     * @author
```

```
    Administrator
 *  @param args
 *  @since JDK 1.6
 */
    public static void main(String[] args){
        new MainFrame();
    }
}
```

项目十五

代码如下:
创建数据库代码

```sql
CREATE DATABASE  HNIUEAMDB CHARACTER SET gbk;
USE HNIUEAMDB;
CREATE TABLE t_material_category (              —— 创建教材类别表
material_id VARCHAR(10) NOT NULL PRIMARY KEY,   —— 教材类别编号
material_name VARCHAR(50) NOT NULL,             —— 教材类别名称
material_memo TEXT NULL                         —— 备注
);
CREATE TABLE t_material_information(
material_information_id VARCHAR(10) NOT NULL PRIMARY KEY,   —— 教材信息编号
material_category_id VARCHAR(10) NOT NULL,      —— 教材类别编号
material_name VARCHAR(50) NOT NULL,             —— 教材名称
material_ISBN VARCHAR(20) NOT NULL,             —— 教材 ISBN 编号
author VARCHAR(20) NOT NULL,                    —— 作者
material_publisher VARCHAR(50) NOT NULL,        —— 出版社
material_price FLOAT NOT NULL,                  —— 价格
material_publication_date datetime NOT NULL     —— 出版时间
);
INSERT INTO t_material_category(material_id,material_name,material_memo)
SELECT 'JSJ001','计算机类','计算机' UNION
SELECT 'JXL002','机械类','机械' UNION
SELECT 'KPL004','科普类','科普' UNION
SELECT 'WXL003','文学类','文学';
INSERT INTO t_material_information
SELECT 'JC001','JSJ001','C 语言程序设计','9786890234','谭浩强','清华大学出版社',32,'2010-04-01' UNION
SELECT 'JC002','JSJ001','数据结构','9786589078','唐森宝','电子工业出版社',28,'2008-07-01' UNION
SELECT 'JC003','WXL003','红楼梦','9786510983','曹雪芹','机械工业出版社',90,'2004-01-01';
```

DBUtil.java 代码

```java
package com.software.dao;
import java.sql.Connection;
import java.sql.DriverManager;
import java.sql.SQLException;
/**
 * 类名：DBUtil <br/>
 * 功能：自定义数据库工具类，封装通用数据库操作 <br/>
 * 创建时间：2016-5-31 下午 4:51:22 <br/>
 * @author Administrator
 * @version
 * @since JDK 1.6
 */
public class DBUtil {
    /**
     * getConnection:连接数据库返回,数据库连接对象. <br/>
     * @author Administrator
     * @return 数据库连接对象
     * @throws ClassNotFoundException
     * @throws SQLException
     * @since JDK 1.6
     */
    public static Connection getConnection() throws ClassNotFoundException,SQLException{
        String url="jdbc:mysql://127.0.0.1:8306/HNIUEAMDB";
        String user="root";
        String pwd="1234";
        Class.forName("com.mysql.jdbc.Driver");
        Connection conn=DriverManager.getConnection(url,user,pwd);
        return conn;
    }
}
```

MyTableModel.java 代码

```java
package com.software.dao;
import java.util.Vector;
import javax.swing.table.DefaultTableModel;
/**
 * 类名：MyTableModel <br/>
 * 功能：自定义 JTable 表格数据模型代码. <br/>
 * 创建时间：2016-5-31 下午 5:03:10 <br/>
 * @author Administrator
 * @version
```

```java
 * @since JDK 1.6
 */
public class MyTableModel extends DefaultTableModel {
    /**
     * 创建一个新的实例 MyTableModel.
     * @param data 表格数据
     * @param columnNames 显示表格的列名
     */
    public MyTableModel(Vector data, Vector columnNames) {
        super(data, columnNames);
    }
    /**
     * 控制表格单元格是否可编辑.
     * @param r 行号
     * @param c 列号
     */
    public boolean isCellEditable(int r, int c) {
        return false;
    }
    /**
     * 获得单元格列类型.         * @param c 列号
     * @return 列类型
     */
            public Class getColumnClass(int c) {
        return getValueAt(0, c).getClass();
    }
}
```

CategoryDao.java 代码

```java
package com.software.dao;
import static com.software.dao.DBUtil.getConnection;
import java.sql.Connection;
import java.sql.ResultSet;
import java.sql.SQLException;
import java.sql.Statement;
import java.util.Vector;
import com.software.entity.CategoryBean;
/**
 * 类名：CategoryDao <br/>
 * 功能：封装了教材分类表的数据操作类 <br/>
 * 创建时间：2016-5-31 下午 7:06:13 <br/>
 * @author Administrator
 * @version
```

```java
 * @since JDK 1.6
 */
public class CategoryDao {
    /**
     * queyCategory:查询教材分类信息.<br/>
     * @author Administrator
     * @return 返回教材分类查询结果
     * @since JDK 1.6
     */
    public static Vector queyCategory(){
        Vector data=new Vector();
        Connection conn=null;
        Statement stm=null;
        ResultSet rs=null;
        String sql="SELECT * FROM t_material_category";
        try {
            conn=getConnection();
            stm=conn.createStatement();
            rs=stm.executeQuery(sql);
            while(rs.next()){
                CategoryBean category=new CategoryBean();
                category.setCategory_id(rs.getString("material_id"));
                category.setCategory_name(rs.getString("material_name"));
                data.add(category);
            }
        } catch (ClassNotFoundException e) {
            e.printStackTrace();
        } catch (SQLException e) {
            System.out.println(sql);
            e.printStackTrace();
        }finally{
            try {
                if(rs!=null) rs.close();
                if(stm!=null)stm.close();
                if(conn!=null) conn.close();
            } catch (SQLException e) {
                e.printStackTrace();
            }
        }
        return data;
    }
}
```

MaterialDao.java 代码

```java
package com.software.dao;
import java.sql.Connection;
import java.sql.PreparedStatement;
import java.sql.ResultSet;
import java.sql.SQLException;
import java.sql.Statement;
import java.util.Vector;
import static com.software.dao.DBUtil.*;
import javax.swing.table.DefaultTableModel;
import javax.swing.table.TableModel;
import com.software.entity.CategoryBean;
import com.software.entity.MaterialBean;
/**
 * 类名：MaterialDao <br/>
 * 功能：封装了教材表的数据库操作. <br/>
 * 创建时间：2016-6-1 上午 8:11:52 <br/>
 * @author Administrator
 * @version
 * @since JDK 1.6
 */
public class MaterialDao {
    /**
     * getColumnNames：获取显示教材信息表格列标题. <br/>
     * @author Administrator
     * @return 教材信息表格列标题
     * @since JDK 1.6
     */
    public static Vector<String> getColumnNames(){
        Vector<String> columnNames=new Vector<String>();
        columnNames.add(0,"教材 ID");
        columnNames.add(1,"教材类别名称");
        columnNames.add(2,"教材名称");
        return columnNames;
    }
    /**
     * getEmpty：获取一个带中文列标题的教材信息空表格数据模型. <br/>
     * @author Administrator
     * @return 中文列标题的教材信息空表格数据模型
     * @since JDK 1.6
```

```java
 */
public static TableModel getEmpty(){
    Vector data=new Vector();
    DefaultTableModel dmt=new DefaultTableModel(data, getColumnNames());
    return dmt;
}
/**
 * queryMaterials:查询教材信息.<br/>
 * @author Administrator
 * @param material_name 教材名称
 * @return 查询教材信息结果
 * @since JDK 1.6
 */
public Vector queryMaterials(String material_name){
    Vector data=new Vector();
    Connection conn=null;
    Statement stm=null;
    ResultSet rs=null;
    String sql="SELECT material_information_id,"
            +"b.material_name as category_name,a.material_name as material_name"
            +" FROM t_material_information a,t_material_category b"
            +" WHERE a.material_category_id=b.material_id";
    try {
                if(!(material_name==null||material_name.equals(""))){
            sql+=" and a.material_name='"+material_name+"'";
        }
        conn=getConnection();
        stm=conn.createStatement();
        rs=stm.executeQuery(sql);
        while(rs.next()){
            Vector row=new Vector();
            row.add(rs.getString(1));
            row.add(rs.getString(2));
            row.add(rs.getString(3));
            data.add(row);
        }
    catch (ClassNotFoundException e) {
        e.printStackTrace();
    } catch (SQLException e) {
        System.out.println(sql);
        e.printStackTrace();
    }finally{
```

```java
            try {
                if(rs!=null) rs.close();
                if(stm!=null) stm.close();
                if(conn!=null) conn.close();
            } catch (SQLException e) {
                e.printStackTrace();
            }
        }
        return data;
    }
    /**
     * querySingleMaterial:依据教材ID查询教材信息。<br/>
     * @author Administrator
     * @param material_id 教材ID
     * @return 教材信息查询结果
     * @since JDK 1.6
     */
    public MaterialBean querySingleMaterial(String material_id){
        MaterialBean material=new MaterialBean();
        Connection conn=null;
        Statement stm=null;
        ResultSet rs=null;
        String sql="SELECT a.*,b.material_name as category_name"
                +" FROM t_material_information a,t_material_category b "
                +" WHERE a.material_category_id=b.material_id and a.material_information_id='"+material_id+"'";
        try {
            conn=getConnection();
            stm=conn.createStatement();
            rs=stm.executeQuery(sql);
            while(rs.next()){
                material.setMaterial_id(rs.getString("material_information_id"));
                material.setMaterial_name(rs.getString("material_name"));
                material.setMaterial_ISBN(rs.getString("material_ISBN"));
                material.setAuthor(rs.getString("author"));
                material.setMaterial_publisher(rs.getString("material_publisher"));
                material.setMaterial_price(rs.getFloat("material_price"));
                material.setMaterial_publication_date(rs.getString("Material_publication_date").substring(0,10));
                material.setCategory(new CategoryBean(rs.getString("material_category_id"),rs.getString("category_name")));
            }
```

```java
            } catch (ClassNotFoundException e) {
                e.printStackTrace();
            } catch (SQLException e) {
                System.out.println(sql);
                e.printStackTrace();
            } finally {
                try {
                    if(rs != null) rs.close();
                    if(stm != null) stm.close();
                    if(conn != null) conn.close();
                } catch (SQLException e) {
                    e.printStackTrace();
                }
            }
            return material;
        }
        /**
         * updateMaterial:修改教材信息 <br/>
         * @author Administrator
         * @param material 教材信息实体对象
         * @return 修改是否成功
         * @since JDK 1.6
         */
        public boolean updateMaterial(MaterialBean material){
            boolean isSuccess = false;
            Connection conn = null;
            PreparedStatement ps = null;
            try {
                String sql = "update t_material_information set material_category_id=?,"
                        + "material_name=?,"
                        + "material_ISBN=?,"
                        + "author=?,"
                        + "material_publisher=?,"
                        + "material_price=?,"
                        + "material_publication_date=? "
                        + " where material_information_id=? ";
                conn = getConnection();
                ps = conn.prepareStatement(sql);
                ps.setString(1, material.getCategory().getCategory_id());
                ps.setString(2, material.getMaterial_name());
                ps.setString(3, material.getMaterial_ISBN());
                ps.setString(4, material.getAuthor());
```

```java
                ps.setString(5, material.getMaterial_publisher());
                ps.setFloat(6, material.getMaterial_price());
                ps.setString(7, material.getMaterial_publication_date());
                ps.setString(8, material.getMaterial_id());
                ps.executeUpdate();
                isSuccess=true;
        } catch (ClassNotFoundException e) {
            e.printStackTrace();
        } catch (SQLException e) {
            e.printStackTrace();
        }finally{
                try {
                    if(ps!=null) ps.close();
                    if(conn!=null) conn.close();
                } catch (SQLException e) {
                            e.printStackTrace();
                }
        }
            return isSuccess;
    }
}
```

CategoryBean.java 代码

```java
package com.software.entity;
/**
 * 类名：CategoryBean <br/>
 * 功能：对数据库中教材分类表(t_material_category)的实体类. <br/>
 * 创建时间：2016-6-1 上午 8:23:46 <br/>
 * @author Administrator
 * @version
 * @since JDK 1.6
 */
public class CategoryBean {
    private String category_id;
        private String category_name;
        public CategoryBean() {
        }
    public CategoryBean(String category_id, String category_name) {
        this.category_id = category_id;
        this.category_name = category_name;
    }
        public String getCategory_id() {
            return category_id;
```

```java
    }
        public void setCategory_id(String category_id) {
            this.category_id = category_id;
        }
        public String getCategory_name() {
            return category_name;
        }
        public void setCategory_name(String category_name) {
            this.category_name = category_name;
        }
        /*
         * 重写 equals 方法用于下拉列表
         */
        @Override
        public boolean equals(Object obj) {
            return ((CategoryBean)obj).getCategory_name().equals(category_name) ;
        }
        /*
         * 重写 toString 方法用于下拉列表
         */
        @Override
        public String toString() {
            return  category_name;
        }
}
```

MaterialBean.java 代码

```java
package com.software.entity;
/**
 * 类名：MaterialBean <br/>
 * 功能：对应数据库中教材信息表(t_material_information)实体类. <br/>
 * 创建时间：2016-6-1 上午 8:27:40 <br/>      *
 * @author Administrator
 * @version
 * @since JDK 1.6
 */
public class MaterialBean {
    private String material_id;
    private CategoryBean category;
    private String material_name;
    private String material_ISBN;
    private String author;
    private String material_publisher;
```

```java
    private float material_price;
    private String material_publication_date;
    public String getMaterial_id() {
        return material_id;
    }
    public void setMaterial_id(String material_id) {
        this.material_id = material_id;
    }
    public CategoryBean getCategory() {
        return category;
    }
    public void setCategory(CategoryBean category) {
        this.category = category;
    }
    public String getMaterial_name() {
        return material_name;
    }
    public void setMaterial_name(String material_name) {
        this.material_name = material_name;
    }
    public String getMaterial_ISBN() {
        return material_ISBN;
    }
    public void setMaterial_ISBN(String material_ISBN) {
        this.material_ISBN = material_ISBN;
    }
    public String getAuthor() {
        return author;
    }
    public void setAuthor(String author) {
        this.author = author;
    }
    public String getMaterial_publisher() {
        return material_publisher;
    }
    public void setMaterial_publisher(String material_publisher) {
        this.material_publisher = material_publisher;
    }
    public float getMaterial_price() {
        return material_price;
    }
    public void setMaterial_price(float material_price) {
        this.material_price = material_price;
```

```java
    }
    public String getMaterial_publication_date() {
        return material_publication_date;
    }
    public void setMaterial_publication_date(String material_publication_date) {
        this.material_publication_date = material_publication_date;
    }
}
```

QueryPanel.java 代码

```java
package com.software.ui;
import java.awt.BorderLayout;
import java.awt.FlowLayout;
import java.awt.event.ActionEvent;
import java.awt.event.ActionListener;
import java.util.Vector;
import javax.swing.BorderFactory;
import javax.swing.JButton;
import javax.swing.JFrame;
import javax.swing.JLabel;
import javax.swing.JOptionPane;
import javax.swing.JPanel;
import javax.swing.JScrollPane;
import javax.swing.JTable;
import javax.swing.JTextField;
import javax.swing.SwingConstants;
import javax.swing.table.DefaultTableModel;
import com.software.dao.MaterialDao;
import com.software.dao.MyTableModel;
import static com.software.dao.MaterialDao.*;
/**
 * 类名：QueryPanel <br/>
 * 功能：查询教材信息界面元素布局面板。<br/>
 * 创建时间：2016-6-1 上午 8:32:12 <br/>
 * @author Administrator
 * @version
 * @since JDK 1.6
 */
public class QueryPanel extends JPanel {
    MainFrame parent;
    JPanel jp_condition=new JPanel();
    JPanel jp_update=new JPanel(new FlowLayout(FlowLayout.RIGHT));
```

```java
    JLabel jlb_name=new JLabel("教材名称：",SwingConstants.RIGHT);
    JTextField jtf_name=new JTextField(30);
    JButton jb_query=new JButton("查询");
    JButton jb_update=new JButton("修改");
    JTable
jtable_query=new JTable();
    JScrollPane jScrollPane=new JScrollPane();
    /**
     * 新建一个查询教材面板实例,对界面控件进行布局
     */
    public QueryPanel(MainFrame parent){
        this.parent=parent;
        setSize(400,300);
        setLayout(new BorderLayout());
            jtable_query.setModel(getEmpty());
        jp_condition.add(jlb_name);
        jp_condition.add(jtf_name);
        jp_condition.add(jb_query);
        jp_condition.setBorder(BorderFactory.createTitledBorder("==查询条件=="));
        jp_update.add(jb_update);
        jScrollPane.setBorder(BorderFactory.createTitledBorder("==教材信息=="));
        jScrollPane.setViewportView(jtable_query);
        add(jp_condition,BorderLayout.NORTH);
        add(jScrollPane);
        add(jp_update,BorderLayout.SOUTH);
        jb_query.addActionListener(new QueryHandler());
        jb_update.addActionListener(new UpdateHandler());
    }
    /**
     * queryMaterials:查询教材信息刷新数据表格。<br/>
     * @author Administrator
     * @since JDK 1.6
     */
    public void queryMaterials(){
        MaterialDao dao=new MaterialDao();
        Vector data=dao.queryMaterials(jtf_name.getText());
        if(data.size()<1){
            JOptionPane.showMessageDialog(null,"没有查询到数据,请更改查询条件!");
        }else{
            MyTableModel mt=new MyTableModel(data,getColumnNames());
            jtable_query.setModel(mt);
        }
```

```java
    }
    /**
     * 查询按钮事件处理类
     */
    class QueryHandler implements ActionListener{
        public void actionPerformed(ActionEvent e) {
            queryMaterials();
        }
    }
    /**
     * 修改按钮事件处理类
     */
    class UpdateHandler implements ActionListener{
        @Override
        public void actionPerformed(ActionEvent e) {
            if(jtable_query.getSelectedRowCount()>0){
                DefaultTableModel tableModel = (DefaultTableModel) jtable_query.getModel();
                String material_id=(String)tableModel.getValueAt(jtable_query.getSelectedRow(),0);
                new MaterialDialog(parent,material_id);
            }else{
                JOptionPane.showMessageDialog(null,"没有选中要修改的记录");
            }
        }
    }
}
```

MaterialDialog.java 代码
```java
package com.software.ui;
import java.awt.Container;
import java.awt.event.ActionEvent;
import java.awt.event.ActionListener;
import javax.swing.JButton;
import javax.swing.JComboBox;
import javax.swing.JDialog;
import javax.swing.JFrame;
import javax.swing.JLabel;
import javax.swing.JOptionPane;
import javax.swing.JTextField;
import javax.swing.SwingConstants;
import com.software.dao.MaterialDao;
import com.software.entity.CategoryBean;
import com.software.entity.MaterialBean;
import com.software.ui.MainFrame;
```

```java
import static com.software.dao.CategoryDao.*;
/**
 * 类名：MaterialDialog <br/>
 * 功能：修改教材信息对话框架窗体类. <br/>
 * 创建时间：2016-6-1 上午 8:39:43 <br/>
 * @author Administrator
 * @version
 * @since JDK 1.6
 */
public class MaterialDialog extends JDialog
    implements ActionListener{
    private MainFrame parent;
    private String material_id;
    JTextField jtf_id=new JTextField(30);
    JComboBox jcb_category;
    JTextField jtf_name=new JTextField(30);
    JTextField jtf_ISBN=new JTextField(30);
    JTextField jtf_author=new JTextField(30);
    JTextField jtf_publisher=new JTextField(30);
    JTextField jtf_price=new JTextField(30);
    JTextField jtf_publication_date=new JTextField(30);
    JButton jb_ok=new JButton("确定");
    JButton jb_cancle=new JButton("取消");
    /**
     * 新建一个 MaterialDialog 实例，对界面控件进行布局
     * @param parent 父窗体对象实例
     * @param material_id 教材 ID 值
     */
    public MaterialDialog(MainFrame parent,String material_id){
        super(parent,"教材信息修改",true);
        this.parent=parent;
        this.material_id=material_id;
        Container container=this.getContentPane();
        container.setLayout(null);
        this.setSize(400,450);
        JLabel jbl_id=new JLabel("教材信息 ID:",SwingConstants.LEFT);
        add(jbl_id);add(jtf_id);
        jtf_id.setEnabled(false);
        jbl_id.setBounds(20,20,160,30);
        jtf_id.setBounds(180,20,200,30);
        JLabel jbl_category=new JLabel("教材类别名称:",SwingConstants.LEFT);
        jcb_category=new JComboBox(queyCategory());
        add(jbl_category);add(jcb_category);
```

```java
        jbl_category.setBounds(20, 60, 160, 30);
        jcb_category.setBounds(180,60,200,30);
        JLabel jbl_name=new JLabel("教材名称:",SwingConstants.LEFT);
        add(jbl_name);add(jtf_name);
        jbl_name.setBounds(20,100, 160, 30);
        jtf_name.setBounds(180,100,200,30);
        JLabel jbl_ISBN=new JLabel("教材ISBN编号:",SwingConstants.LEFT);
        add(jbl_ISBN);
        add(jtf_ISBN);
        jbl_ISBN.setBounds(20,140, 160, 30);
        jtf_ISBN.setBounds(180,140,200,30);
        JLabel jbl_author=new JLabel("作者:",SwingConstants.LEFT);
        add(jbl_author);add(jtf_author);
        jbl_author.setBounds(20,180, 160, 30);
        jtf_author.setBounds(180,180,200,30);
        JLabel jbl_publisher=new JLabel("出版社:",SwingConstants.LEFT);
        add(jbl_publisher);add(jtf_publisher);
        jbl_publisher.setBounds(20,220, 160, 30);
        jtf_publisher.setBounds(180,220,200,30);
        JLabel jbl_price=new JLabel("价格:",SwingConstants.LEFT);
        add(jbl_price);add(jtf_price);
        jbl_price.setBounds(20,260, 160, 30);
        jtf_price.setBounds(180,260,200,30);
        JLabel jbl_publication_date=new JLabel("出版时间(YYYY-MM-DD):",SwingConstants.LEFT);
        add(jbl_publication_date);add(jtf_publication_date);
        jbl_publication_date.setBounds(20,300, 160, 30);
        jtf_publication_date.setBounds(180,300,200,30);
        add(jb_ok);add(jb_cancle);
        jb_ok.setBounds(20,340, 80, 30);
        jb_cancle.setBounds(180,340,80,30);
        jb_ok.addActionListener(this);
        jb_cancle.addActionListener(this);
        initData();
        this.setVisible(true);
    }
    /**
     * initData:初始化界面控件数据 <br/>
     * @author     Administrator
     * @since JDK 1.6
     */
    public void initData(){
        MaterialDao materialDao=new MaterialDao();
```

```java
        MaterialBean material=materialDao.querySingleMaterial(material_id);
        jtf_id.setText(material.getMaterial_id());
        jcb_category.setSelectedItem(material.getCategory());
        jtf_name.setText(material.getMaterial_name());
        jtf_ISBN.setText(material.getMaterial_ISBN());
        jtf_author.setText(material.getAuthor());
        jtf_publisher.setText(material.getMaterial_publisher());
        jtf_price.setText(material.getMaterial_price()+"");
        jtf_publication_date.setText(material.getMaterial_publication_date());
    }
    /**
     * updateMaterial:修改教材信息.<br/>
     * @author Administrator
     * @return 是否修改成功
     * @since JDK 1.6
     */
    public boolean updateMaterial(){
        MaterialDao materialDao=new MaterialDao();
        MaterialBean material=new MaterialBean();
        material.setMaterial_id(jtf_id.getText());
        material.setCategory((CategoryBean)jcb_category.getSelectedItem());
        material.setMaterial_name(jtf_name.getText());
        material.setMaterial_ISBN(jtf_ISBN.getText());
        material.setAuthor(jtf_author.getText());
        material.setMaterial_publisher(jtf_publisher.getText());
        material.setMaterial_price(Float.parseFloat(jtf_price.getText()));
        material.setMaterial_publication_date(jtf_publication_date.getText());
        return materialDao.updateMaterial(material);
    }
    /**
     * 按钮事件处理
     */
    @Override
    public void actionPerformed(ActionEvent e) {
        String cmd=e.getActionCommand();
        if (cmd.equals("确定")){
            if(updateMaterial()){
                JOptionPane.showMessageDialog(null,"修改成功");
                parent.queryPanel.queryMaterials();
            }
            else{
                JOptionPane.showMessageDialog(null,"修改不成功请检查数据");
```

```
        }
    }else if(cmd.equals("取消")){
        this.dispose();
    }
}
```

}

MainFrame.java 代码
```
package com.software.ui;
import java.awt.Container;
import javax.swing.JFrame;
import javax.swing.JTabbedPane;
/**
 * 类名：MainFrame <br/>
 * 功能：程序运行主窗体类. <br/>
 * 创建时间：2016-6-1 上午 8:29:43 <br/>
 * @author Administrator
 * @version
 * @since JDK 1.6
 */
public class MainFrame extends JFrame {
    public QueryPanel queryPanel;
    /**
     * 创建一个新的实例 MainFrame,对面板进行界面布局.
     */
    public MainFrame(){
        queryPanel=new QueryPanel(this);
        Container content=this.getContentPane();
        content.add(queryPanel);
        this.setTitle("教材信息查询");
        this.setSize(550,400);
        this.setVisible(true);
        this.setDefaultCloseOperation(JFrame.EXIT_ON_CLOSE);
    }
    /**
     * main:程序运行主函数. <br/>
     * @author Administrator
     * @param args
     * @since JDK 1.6
     */
    public static void main(String[] args){
```

```
        new MainFrame();
    }
}
```

项目十六

代码如下：
创建数据库代码：
CREATE DATABASE HNIUEAMDB CHARACTER SET gbk;
USE HNIUEAMDB;
CREATE TABLE t_department(
    department_id VARCHAR(10) NOT NULL PRIMARY KEY, —— 教材部门编号
    department_name VARCHAR(50) NOT NULL,   —— 部门名称
    department_memo TEXT NULL   —— 备注
);
INSERT INTO t_department
SELECT 'XY0001','信息工程系','信息' UNION
SELECT 'XY0002','机电工程系','机电' UNION
SELECT 'XY0003','计算机工程系','计算机' UNION
SELECT 'XY0004','经济管理系','经济';
CREATE TABLE t_teacher_information(
    teacher_id VARCHAR(10) NOT NULL PRIMARY KEY, —— 教师信息编号
    department_id VARCHAR(10) NOT NULL,   —— 部门编号
    teacher_name VARCHAR(8) NOT NULL,   —— 教师名称
    sex INT NOT NULL,   —— 性别
    age INT NOT NULL,   —— 年龄
    prade VARCHAR(10) NOT NULL   —— 职称
)
INSERT INTO t_teacher_information
SELECT 'YT10001','XY0001','王枚',0,30,'讲师' UNION
SELECT 'YT10002','XY0001','张芳',0,28,'讲师' UNION
SELECT 'YT20001','XY0002','李利',1,45,'教授'

DBUtil.java 代码
package com.software.dao;
import java.sql.Connection;
import java.sql.DriverManager;
import java.sql.SQLException;
/**
 * 类名：DBUtil
 * 功能：自定义数据库工具类，封装通用数据库操作

* 创建时间：2016-5-31 下午 4:51:22
 * @author Administrator
 * @version
 * @since JDK 1.6
 */
public class DBUtil {
    /**
     * getConnection：连接数据库返回，数据库连接对象。
     * @author Administrator
     * @return 数据库连接对象
     * @throws ClassNotFoundException
     * @throws SQLException
     * @since JDK 1.6
     */
    public static Connection getConnection() throws ClassNotFoundException,SQLException{
        String url="jdbc:mysql://127.0.0.1:8306/HNIUEAMDB";
        String user="root";
        String pwd="1234";
        Class.forName("com.mysql.jdbc.Driver");
        Connection conn=DriverManager.getConnection(url, user, pwd);
        return conn;
    }
}

MyTableModel.java 代码
package com.software.dao;
import java.util.Vector;
import javax.swing.table.DefaultTableModel;
/**
 * 类名：MyTableModel
 * 功能：自定义 JTable 表格数据模型代码.
 * 创建时间：2016-5-31 下午 5:03:10
 * @author Administrator
 * @version
 * @since JDK 1.6
 */
public class MyTableModel extends DefaultTableModel {
    /**
     * 创建一个新的实例 MyTableModel.
     * @param data 表格数据
     * @param columnNames 显示表格的列名

```java
 */
public MyTableModel(Vector data, Vector columnNames) {
    super(data, columnNames);
}
/**
 * 控制表格单元格是否可编辑.
 * @param r 行号
 * @param c 列号
 */
public boolean isCellEditable(int r, int c) {
    return false;
}
/**
 * 获得单元格列类型.
 * @param c 列号
 * @return 列类型
 */
public Class getColumnClass(int c) {
    return getValueAt(0, c).getClass();
}
}
```

TeacherDao.java 代码

```java
package com.software.dao;
import java.sql.Connection;
import java.sql.ResultSet;
import java.sql.Statement;
import java.util.Vector;
import javax.swing.table.DefaultTableModel;
import javax.swing.table.TableModel;
import com.software.entity.DepartmentBean;
import com.software.entity.TeacherBean;
import import static com.software.dao.DBUtil.*;
/**
 * 类名：TeacherDao <br/>
 * 功能：封装对教师信息表的数据库操作. <br/>
 * 创建时间：2016-6-1 上午 9:54:39 <br/>
 * @author Administrator
 * @version
 * @since JDK 1.6
 */
public class TeacherDao {
    /**
```

```java
 * getColumnNames:获取显示教师信息表格列标题. <br/>
 * @author Administrator
 * @return 教师信息表格列标题
 * @since JDK 1.6
 */
public static Vector<String> getColumnNames() {
    Vector<String> columnNames = new Vector<String>();
    columnNames.add(0,"教师编号");
    columnNames.add(1,"部门名称");
    columnNames.add(2,"教师姓名");
    return columnNames;
}
/**
 * getEmpty:获取一个带中文列标题的教师信息空表格数据模型. <br/>
 * @author Administrator
 * @return 中文列标题的教师信息空表格数据模型
 * @since JDK 1.6
 */
public static TableModel getEmpty() {
    Vector data = new Vector();
    DefaultTableModel dmt = new DefaultTableModel(data, getColumnNames());
    return dmt;
}
/**
 * queryTeachers:查询教师信息. <br/>
 * @author Administrator
 * @param teacher_name 教师姓名
 * @return 教师信息查询结果
 * @since JDK 1.6
 */
public Vector queryTeachers(String teacher_name) {
    Vector data = new Vector();
    Connection conn = null;
    Statement stm = null;
    ResultSet rs = null;
    String sql = "SELECT a.*,b.department_name FROM t_teacher_information a,t_department b where a.department_id=b.department_id";
    try {
        if (!(teacher_name == null || teacher_name.equals(""))) {
            sql = sql + " and teacher_name='" + teacher_name + "'";
        }
```

```java
            conn = getConnection();
            stm = conn.createStatement();
            rs = stm.executeQuery(sql);
            while (rs.next()) {
                Vector row = new Vector();
                row.add(rs.getString("teacher_id"));
                row.add(rs.getString("teacher_name"));
                row.add(rs.getString("department_name"));
                data.add(row);
            }
        } catch (Exception e) {
            System.out.println(sql);
            e.printStackTrace();
        } finally {
            try {
                if (rs != null)
                    rs.close();
                if (stm != null)
                    stm.close();
                if (conn != null)
                    conn.close();
            } catch (Exception e) {
                e.printStackTrace();
            }
        }
        return data;
    }

    /**
     * querySingleTeacher:依据教师ID查询教师信息.<br/>
     * @author Administrator
     * @param teacher_id 教师ID
     * @return 教师信息查询结果
     * @since JDK 1.6
     */
    public TeacherBean querySingleTeacher(String teacher_id) {
        TeacherBean teacher = new TeacherBean();
        Connection conn = null;
        Statement stm = null;
        ResultSet rs = null;
        String sql = "SELECT a.*,b.department_name FROM t_teacher_information a,t_department b
```

```java
        where a.department_id=b.department_id "
                + " and a.teacher_id='" + teacher_id + "'";
        try {
            conn = getConnection();
            stm = conn.createStatement();
            rs = stm.executeQuery(sql);
            while (rs.next()) {
                teacher.setTeacher_id(rs.getString("teacher_id"));
                teacher.setTeacher_name(rs.getString("teacher_name"));
                teacher.setSex(rs.getInt("sex"));
                teacher.setAge(rs.getInt("age"));
                teacher.setPrade(rs.getString("prade"));
                DepartmentBean department = new DepartmentBean(rs.getString("department_id"),
                        rs.getString("department_name"));
                teacher.setDepartment(department);
            }
        } catch (Exception e) {
            System.out.println(sql);
            e.printStackTrace();
        } finally {
            try {
                if (rs != null)
                    rs.close();
                if (stm != null)
                    stm.close();
                if (conn != null)
                    conn.close();
            } catch (Exception e) {

                e.printStackTrace();
            }
        }
        return teacher;
    }
    /**
     * deleteTeacher:依据教师 ID 删除教师信息. <br/>
     * @author      Administrator
     * @param teacher_id 教师 ID
     * @return 删除记录的行数
     * @since JDK 1.6
     */
    public int deleteTeacher(String teacher_id) {
        Connection conn = null;
```

```java
        Statement stm = null;
        int rowCount = 0;
        String sql = "delete from t_teacher_information where teacher_id='" + teacher_id + "'";
        try {
            conn = getConnection();
            stm = conn.createStatement();
            rowCount = stm.executeUpdate(sql);
        } catch (Exception e) {
            System.out.println(sql);
            e.printStackTrace();
        } finally {
            try {
                if (stm != null)
                    stm.close();
                if (conn != null)
                    conn.close();
            } catch (Exception e) {
                e.printStackTrace();
            }
        }
        return rowCount;
    }
}
```

DepartmentBean.java 代码

```java
package com.software.entity;
/**
 * 类名：DepartmentBean <br/>
 * 功能：对应数据库中部门信息表(t_department)实体类。<br/>
 * 创建时间：2016-6-1 上午10:18:34 <br/>
 * @author Administrator
 * @version
 * @since JDK 1.6
 */
public class DepartmentBean {
    private String department_id;
    private String department_name;
    public DepartmentBean(String department_id, String department_name) {
        super();
        this.department_id = department_id;
        this.department_name = department_name;
    }
    public String getDepartment_id() {
```

```java
        return department_id;
    }
    public void setDepartment_id(String department_id) {
        this.department_id = department_id;
    }
    public String getDepartment_name() {
        return department_name;
    }
    public void setDepartment_name(String department_name) {
        this.department_name = department_name;
    }

}
```

TeacherBean.java 代码

```java
package com.software.entity;
/**
 * 类名：TeacherBean <br/>
 * 功能：对应数据库中教师信息表(t_teacher_information)实体类. <br/>
 * 创建时间：2016-6-1 上午 10:19:55 <br/>
 * @author Administrator
 * @version
 * @since JDK 1.6
 */
public class TeacherBean {
    private String teacher_id;
    private DepartmentBean department;
    private String teacher_name;
    private int sex;
    private int age;
    private String prade;
    public String getTeacher_id() {
        return teacher_id;
    }
    public void setTeacher_id(String teacher_id) {
        this.teacher_id = teacher_id;
    }
    public DepartmentBean getDepartment() {
        return department;
    }
    public void setDepartment(DepartmentBean department) {
        this.department = department;
```

```java
    }
    public String getTeacher_name() {
        return teacher_name;
    }
    public void setTeacher_name(String teacher_name) {
        this.teacher_name = teacher_name;
    }
    public int getSex() {
        return sex;
    }
    public void setSex(int sex) {
        this.sex = sex;
    }
    public int getAge() {
        return age;
    }
    public void setAge(int age) {
        this.age = age;
    }
    public String getPrade() {
        return prade;
    }
    public void setPrade(String prade) {
        this.prade = prade;
    }
}
```

TeacherDialog .java 代码

```
package com.software.ui;
import java.awt.Container;
import java.awt.event.ActionEvent;
import java.awt.event.ActionListener;
import javax.swing.BorderFactory;
import javax.swing.JButton;
import javax.swing.JDialog;
import javax.swing.JFrame;
import javax.swing.JLabel;
import javax.swing.JPanel;
import com.software.dao.TeacherDao;
import com.software.entity.TeacherBean;
/**
 * 类名：TeacherDialog <br/>
```

* 功能：显示教师详细信息的对话框窗体类.
 * 创建时间：2016-6-1 上午 10:32:29
           *
 * @author Administrator
 * @version
 * @since JDK 1.6
 */
public class TeacherDialog extends JDialog implements ActionListener{
    /**
     * 创建一个新的实例 TeacherDialog，对界面控件布局并初始化数据.
     * @param parent 打开对话框的窗体实例
     * @param teacher_id 教师 ID
     */
    public TeacherDialog(JFrame parent,String teacher_id){
        super(parent,"教师信息",true);
        TeacherDao teacherDao=new TeacherDao();
        TeacherBean teacher=teacherDao.querySingleTeacher(teacher_id);
        JLabel jbl_teacher_id=new JLabel("教师编号:\t"+teacher.getTeacher_id());
        JLabel jbl_department_name=new JLabel("部门名称:\t"+teacher.getDepartment().getDepartment_name());
        JLabel jbl_teacher_name=new JLabel("教师姓名:\t"+teacher.getTeacher_name());
        JLabel jbl_teacher_sex;
        if(teacher.getSex()==1)
            jbl_teacher_sex=new JLabel("教师性别:\t 男");
        else
            jbl_teacher_sex=new JLabel("教师性别:\t 女");
        JLabel jbl_teacher_age=new JLabel("教师年龄:\t"+teacher.getAge());
        JLabel jbl_teacher_prade=new JLabel("教师职称:\t"+teacher.getPrade());
        JButton jb_close=new JButton("关闭");
        JPanel jp_detail=new JPanel();
        jp_detail.setLayout(null);
        getContentPane().add(jp_detail);
        jp_detail.setBorder(BorderFactory.createTitledBorder("==教师详细信息=="));
        jp_detail.add(jbl_teacher_id);
        jbl_teacher_id.setBounds(20,20,200,30);
        jp_detail.add(jbl_department_name);
        jbl_department_name.setBounds(20,60,200,30);
        jp_detail.add(jbl_teacher_name);
        jbl_teacher_name.setBounds(20,100,200,30);
        jp_detail.add(jbl_teacher_sex);
        jbl_teacher_sex.setBounds(20,140,200,30);
        jp_detail.add(jbl_teacher_age);
        jbl_teacher_age.setBounds(20,180,200,30);
        jp_detail.add(jbl_teacher_prade);

```java
            jbl_teacher_prade.setBounds(20, 220, 200, 30);
            jp_detail.add(jb_close);
            jb_close.setBounds(100, 260, 80, 30);
            jb_close.addActionListener(this);
            this.setSize(300, 350);
            this.setVisible(true);
    }
    /*
     * 按钮事件处理
     */
    public void actionPerformed(ActionEvent e) {
            this.dispose();
    }
}
```

MainFrame.java 代码

```java
package com.software.ui;
import java.awt.BorderLayout;
import java.awt.Container;
import java.awt.FlowLayout;
import java.awt.event.ActionEvent;
import java.awt.event.ActionListener;
import java.util.Vector;
import javax.swing.BorderFactory;
import javax.swing.JButton;
import javax.swing.JFrame;
import javax.swing.JLabel;
import javax.swing.JOptionPane;
import javax.swing.JPanel;
import javax.swing.JScrollPane;
import javax.swing.JTable;
import javax.swing.JTextField;
import javax.swing.SwingConstants;
import javax.swing.table.DefaultTableModel;
import com.software.dao.MyTableModel;
import com.software.dao.TeacherDao;
import static com.software.dao.TeacherDao.*;
/**
 * 类名：MainFrame <br/>
 * 功能：程序运行主窗体类。<br/>
 * 创建时间：2016-6-1 上午 10:21:18 <br/>
 * @author Administrator
```

```java
 * @version
 * @since JDK 1.6
 */
public class MainFrame extends JFrame implements ActionListener {
    JPanel jp_condition = new JPanel();
    JLabel jlb_name = new JLabel("教师姓名:", SwingConstants.RIGHT);
    JTextField jtf_name = new JTextField(30);
    JButton jb_query = new JButton("查询");
    JTable jtable_query = new JTable();
    JScrollPane jsp = new JScrollPane();
    JPanel jp_south = new JPanel(new FlowLayout(FlowLayout.RIGHT));
    JButton jb_delete = new JButton("删除");
    JButton jb_detail = new JButton("查看详细信息");
    /**
     * 创建一个新的实例 MainFrame. 对界面控件进行布局
     */
    public MainFrame(){
        jp_condition.add(jlb_name);
        jp_condition.add(jtf_name);
        jp_condition.add(jb_query);
        jp_condition.setBorder(BorderFactory.createTitledBorder("教师信息查询"));
        jtable_query.setModel(getEmpty());
        jsp.setViewportView(jtable_query);
        jsp.setBorder(BorderFactory.createTitledBorder("教师信息"));
        jp_south.add(jb_delete);
        jp_south.add(jb_detail);
        Container c = this.getContentPane();
        c.add(jp_condition, BorderLayout.NORTH);
        c.add(jsp);
        c.add(jp_south, BorderLayout.SOUTH);
        jb_query.addActionListener(this);
        jb_delete.addActionListener(this);
        jb_detail.addActionListener(this);
        this.setTitle("教师信息管理");
        this.setSize(550, 400);
        this.setVisible(true);
        this.setDefaultCloseOperation(JFrame.EXIT_ON_CLOSE);
    }
    /**
     * queryTeachers:查询教师信息,刷新表格数据. <br/>
     * @author
Administrator
     * @since JDK 1.6
```

```java
 */
public void queryTeachers(){
    TeacherDao teacherDao=new TeacherDao();
    Vector data=teacherDao.queryTeachers(jtf_name.getText());
    if(data.size()<1){
        JOptionPane.showMessageDialog(null,"没有查询到数据,请更改查询条件!");
    }else{
        MyTableModel mt=new MyTableModel(data,getColumnNames());
        jtable_query.setModel(mt);
    }
}
/**
 * showDetail:创建对话框实例,显示教师详细信息.<br/>
 * @author    Administrator
 * @since JDK 1.6
 */
public void showDetail(){
    if(jtable_query.getSelectedRowCount()>0){
        DefaultTableModel tableModel = (DefaultTableModel) jtable_query.getModel();
        String teacher_id=(String)tableModel.getValueAt(jtable_query.getSelectedRow(),0);
        new TeacherDialog(this,teacher_id);
    }else{
        JOptionPane.showMessageDialog(null,"没有选中要查看的记录");
    }
}
/**
 * deleteTeacher:删除表格中选中的教师信息.<br/>
 * @author
Administrator
 * @since JDK 1.6
 */
public void deleteTeacher(){
    if(jtable_query.getSelectedRowCount()>0){
        DefaultTableModel tableModel = (DefaultTableModel) jtable_query.getModel();
        String teacher_id=(String)tableModel.getValueAt(jtable_query.getSelectedRow(),0);
        int answer=JOptionPane.showConfirmDialog(this,"您确认要删除编号["+teacher_id+"]教师记录吗?","删除确认",JOptionPane.YES_NO_OPTION,JOptionPane.QUESTION_MESSAGE);
        if(answer==0){
            TeacherDao teacherDao=new TeacherDao();
            teacherDao.deleteTeacher(teacher_id);
            queryTeachers();
        }
    }else{
```

```java
                JOptionPane.showMessageDialog(null,"没有选中要删除的记录");
        }
    }
    /**
     * main:主函数程序运行入口. <br/>
     * @author      Administrator
     * @param args
     * @since JDK 1.6
     */
    public static void main(String[] args) {
        new MainFrame();
    }
    /*
     * 按钮事件处理
     */
    @Override
    public void actionPerformed(ActionEvent e) {
        if(e.getActionCommand().equals("查询")){
            queryTeachers();
        }else if(e.getActionCommand().equals("删除")){
            deleteTeacher();
        }else{
            showDetail();
        }
    }
}
```

项目十七

代码如下：
创建数据库代码

```sql
CREATE DATABASE DormDB CHARACTER        ——创建数据库
SET gbk；
USE DormDB；       ——切换到当前数据库
CREATE TABLE t_user(
    user_id int NOT NULL PRIMARY KEY,        ——用户编号
    user_name VARCHAR(20) NOT NULL,           ——登录名
    user_password VARCHAR(12) NOT NULL,       ——密码
    user_role char(1) NOT NULL                ——角色
);
INSERT INTO t_user
SELECT 1,'admin','admin','1';
```

```sql
CREATE TABLE t_dormitory_building(
    dormitory_building_id INT NOT NULL PRIMARY KEY,      -- 宿舍楼编号
    dormitory_building_name VARCHAR(64) NOT NULL,        -- 宿舍楼名称
    bed_number int NOT NULL                              -- 本栋楼的寝室的床位数
);
INSERT INTO t_dormitory_building
SELECT 1,'1栋',10 UNION
SELECT 2,'2栋',6;

CREATE TABLE t_room(
    room_id
    INT NOT NULL PRIMARY KEY,              -- 寝室编号
    room_name VARCHAR(64) NOT NULL,        -- 寝室名称
    living_number INT NOT NULL,            -- 已住人数
    dormitory_building_id INT NOT NULL     -- 所属宿舍楼编号
);
ALTER TABLE t_room ADD CONSTRAINT FK_t_room
FOREIGN KEY(dormitory_building_id) REFERENCES t_dormitory_building(dormitory_building_id);
INSERT INTO t_room
SELECT 1,'1-101',3,1 UNION
SELECT 2,'2-202',6,2 UNION
SELECT 3,'2-203',2,2;
```

DBUtil.java 代码

```java
package com.software.dao;
import java.sql.Connection;
import java.sql.DriverManager;
import java.sql.SQLException;
/**
 * 类名：DBUtil <br/>
 * 功能：自定义数据库工具类，封装通用数据库操作 <br/>
 * 创建时间：2016-5-31 下午 4:51:22 <br/>
 * @author Administrator
 * @version
 * @since JDK 1.6
 */
public class DBUtil {
    /**
     * getConnection：连接数据库返回，数据库连接对象. <br/>
     * @author    Administrator
     * @return 数据库连接对象
     * @throws ClassNotFoundException
     * @throws SQLException
```

```java
 * @since JDK 1.6
 */
public static Connection getConnection() throws ClassNotFoundException,SQLException{
    String url="jdbc:mysql://127.0.0.1:8306/DormDB";
    String user="root";
    String pwd="1234";
    Class.forName("com.mysql.jdbc.Driver");
    Connection conn=DriverManager.getConnection(url,user,pwd);
    return conn;
}

}
```

MyTableModel.java 代码

```java
package com.software.dao;
import java.util.Vector;
import javax.swing.table.DefaultTableModel;
/**
 * 类名:MyTableModel <br/>
 * 功能:自定义 JTable 表格数据模型代码. <br/>
 * 创建时间:2016-5-31 下午 5:03:10 <br/>
 * @author Administrator
 * @version
 * @since JDK 1.6
 */
public class MyTableModel extends DefaultTableModel {
    /**
     * 创建一个新的实例 MyTableModel.
     * @param data 表格数据
     * @param columnNames 显示表格的列名
     */
    public MyTableModel(Vector data, Vector columnNames) {
        super(data,columnNames);
    }
    /**
     * 控制表格单元格是否可编辑.
     * @param r 行号
     * @param c 列号
     */
    public boolean isCellEditable(int r, int c) {
        return false;
    }
    /**
```

```
    * 获得单元格列类型.
    * @param c 列号
    * @return 列类型
    */
    public Class getColumnClass(int c) {
        return getValueAt(0, c).getClass();
    }
}

UserDao.java 代码
package com.software.dao;
import java.sql.Connection;
import java.sql.ResultSet;
import java.sql.Statement;
import com.software.entity.UserBean;
import static com.software.dao.DBUtil.*;
/**
 * 类名：UserDao <br/>
 * 功能：封装对用户表数据操作类. <br/>
 * 创建时间：2016-6-1 上午 11:03:01 <br/>
 * @author Administrator
 * @version
 * @since JDK 1.6
 */
public class UserDao {
    /**
     * chcekLogin:检查登录信息. <br/>
     * @author     Administrator
     * @param user 用户表实体类的实例
     * @since JDK 1.6
     */
    public void chcekLogin(UserBean user){
        Connection conn=null;
        Statement stm=null;
        ResultSet rs=null;
        String sql="SELECT * FROM t_user where user_name='"+user.getUser_name()+"'";
        try {
            conn=getConnection();
            stm=conn.createStatement();
            rs=stm.executeQuery(sql);
            user.setErr_message("当前用户名不存在!");
            while(rs.next()){
                String user_password=rs.getString("user_password");
```

```java
                if(user_password.equals(user.getUser_password())){
                    user.setErr_message(null);
                }else{
                    user.setErr_message("密码错误!");
                }
            }
        } catch (Exception e) {
            System.out.println(sql);
            user.setErr_message("数据库操作出错!");
            e.printStackTrace();
        }finally{
            try{
                if(rs!=null) rs.close();
                if(stm!=null) stm.close();
                if(conn!=null) conn.close();
            }catch(Exception e){
                e.printStackTrace();
            }
        }
    }
}
```

DormInfoDao.java 代码

```java
package com.software.dao;
import java.sql.Connection;
import java.sql.ResultSet;
import java.sql.Statement;
import java.util.Vector;
import static com.software.dao.DBUtil.*;
/**
 * 类名：DormInfoDao <br/>
 * 功能：封装寝室信息对数据的操作类.<br/>
 * 创建时间：2016-6-1 上午 11:07:19 <br/>
 * @author Administrator
 * @version
 * @since JDK 1.6
 */
public class DormInfoDao {
    /**
     * getColumnNames:获取显示寝室信息表格列标题.<br/>
     * @author Administrator
     * @return 寝室信息表格列标题
     * @since JDK 1.6
```

```java
    */
    public Vector<String> getColumnNames(){
        Vector<String> columnNames=new Vector<String>();
        columnNames.add(0,"编号");
        columnNames.add(1,"已住人数");
        columnNames.add(2,"可住人数");
        columnNames.add(3,"寝室名称");
        columnNames.add(4,"所属栋数");
        return columnNames;
    }
    /**
    * queryDorimInfo:查询寝室信息.<br/>
    * @author       Administrator
    * @param fullFlag 值为1查询未住满的寝室,其它值时查询所有寝室
    * @return 寝室信息查询结果
    * @since JDK 1.6
    */
    public Vector queryDorimInfo(int fullFlag){
        Vector data = new Vector();
        Connection conn=null;
        Statement stm=null;
        ResultSet rs=null;
        String sql="SELECT a.*,b.dormitory_building_name,b.bed_number "
                  +" FROM t_room as a, t_dormitory_building as b where a.dormitory_building_id=b.dormitory_building_id";
        try {
            if(fullFlag==1){
                sql=sql+" and living_number<bed_number";
            }
            conn=getConnection();
            stm=conn.createStatement();
            rs=stm.executeQuery(sql);
            while(rs.next()){
                Vector row = new Vector();
                row.add(rs.getString("room_id"));
                row.add(rs.getString("living_number"));
                row.add(rs.getString("bed_number"));
                row.add(rs.getString("room_name"));
                row.add(rs.getString("dormitory_building_name"));
                data.add(row);
            }
        } catch (Exception e) {
            System.out.println(sql);
```

```java
                e.printStackTrace();
            }finally{
                try{
                    if(rs!=null) rs.close();
                    if(stm!=null) stm.close();
                    if(conn!=null) conn.close();
                }catch(Exception e){
                    e.printStackTrace();
                }
            }
            return data;
        }
    }
```

UserBean.java 代码

```java
package com.software.entity;
/**
 * 类名：UserBean <br/>
 * 功能：对应数据库中用户信息表(t_user)实体类. <br/>
 * 创建时间：2016-6-1 上午 11:13:04 <br/>
 * @author Administrator
 * @version
 * @since JDK 1.6
 */
public class UserBean {
    private String user_name;
    private String user_password;
    private String err_message;
    public String getUser_name() {
        return user_name;
    }
    public void setUser_name(String user_name) {
        this.user_name = user_name;
    }
    public String getUser_password() {
        return user_password;
    }
    public void setUser_password(String user_password) {
        this.user_password = user_password;
    }
    public String getErr_message() {
        return err_message;
    }
```

```java
    public void setErr_message(String err_message) {
        this.err_message = err_message;
    }

}
```

DormInfoFrame.java 代码

```java
package com.software.ui;
import java.awt.BorderLayout;
import java.awt.Container;
import java.awt.event.ActionEvent;
import java.awt.event.ActionListener;
import java.awt.event.WindowAdapter;
import java.awt.event.WindowEvent;
import java.util.Vector;
import javax.swing.JButton;
import javax.swing.JFrame;
import javax.swing.JPanel;
import javax.swing.JScrollPane;
import javax.swing.JTable;
import com.software.dao.DormInfoDao;
import com.software.dao.MyTableModel;
/**
 * 类名：DormInfoFrame <br/>
 * 功能：查询寝室信息的窗体类. <br/>
 * 创建时间：2016-6-1 上午 11:15:25 <br/>
 * @author Administrator
 * @version
 * @since JDK 1.6
 */
public class DormInfoFrame extends JFrame implements ActionListener {
    JFrame parent;
    JPanel jp_top=new JPanel();
    JButton jb_notFull=new JButton("查看未住满的寝室");
    JButton jb_all=new JButton("查看所有寝室");
    JButton jb_assign=new JButton("寝室分配");
    JTable jtable=new JTable();
    /**
     * 创建一个新的实例 DormInfoFrame,对界面控件进行布局.
     * @param parent 主窗体类的实例
     */
    public DormInfoFrame(JFrame parent){
        this.parent=parent;
```

```java
        jp_top.add(jb_notFull);
        jp_top.add(jb_all);
        jp_top.add(jb_assign);
        JScrollPane jsp=new JScrollPane();
        jsp.setViewportView(jtable);
        Container c=getContentPane();
        c.add(jp_top,BorderLayout.NORTH);
        c.add(jsp);
        jb_notFull.addActionListener(this);
        jb_all.addActionListener(this);
        this.setTitle("DormInfoMain");
        this.setSize(400,300);
        this.setVisible(true);
        this.setLocationRelativeTo(parent);
        this.addWindowListener(new WindowAdapter(){
            public void windowClosing(WindowEvent e) {
                DormInfoFrame.this.dispose();
            }
        });
        queryDormInfo(0);
    }
    /**
     * queryDormInfo:查询寝室信息,并刷新数据表格。<br/>
     * @author      Administrator
     * @param fullFlag
     * @since JDK 1.6
     */
    public void queryDormInfo(int fullFlag){
        DormInfoDao dormInfoDao=new DormInfoDao();
        Vector data=dormInfoDao.queryDorimInfo(fullFlag);
        Vector columnNames=dormInfoDao.getColumnNames();
        MyTableModel tableModel=new MyTableModel(data,columnNames);
        jtable.setModel(tableModel);
    }

    /*
     * 按钮事件处理
     */
    @Override
    public void actionPerformed(ActionEvent e) {
        String cmd=e.getActionCommand();
        if(cmd.equals("查看未住满的寝室")){
            queryDormInfo(1);
```

```java
        }else if(cmd.equals("查看所有寝室")){
            queryDormInfo(0);
        }
    }
}
```

LoginForm.java 代码

```java
package com.software.ui;
import java.awt.Color;
import java.awt.Container;
import java.awt.event.ActionEvent;
import java.awt.event.ActionListener;
import javax.swing.DefaultComboBoxModel;
import javax.swing.JButton;
import javax.swing.JComboBox;
import javax.swing.JFrame;
import javax.swing.JLabel;
import javax.swing.JPasswordField;
import javax.swing.JTextField;
import javax.swing.SwingConstants;
import com.software.dao.UserDao;
import com.software.entity.UserBean;
/**
 * 类名：LoginForm <br/>
 * 功能：登录窗体类. <br/>
 * 创建时间：2016-6-1 上午 11:23:51 <br/>
 * @author Administrator
 * @version
 * @since JDK 1.6
 */
public class LoginForm extends JFrame implements ActionListener{
    JLabel jbl_user_name=new JLabel("用户名:",SwingConstants.RIGHT);
    JLabel jbl_user_password=new JLabel("密码:",SwingConstants.RIGHT);
    JLabel jbl_user_role=new JLabel("角色:",SwingConstants.RIGHT);
    JLabel jbl_error=new JLabel("",SwingConstants.RIGHT);
    JTextField jtf_user_name=new JTextField(30);
    JPasswordField jpf_user_password=new JPasswordField(30);
    JComboBox jcb_user_role=new JComboBox();
    JButton jb_login=new JButton("登录");
    JButton jb_reset=new JButton("重置");
    /**
     * 创建一个新的实例 LoginForm,对界面控件进行布局.
     */
```

```java
public LoginForm(){
    jcb_user_role.setModel(new DefaultComboBoxModel(new String[]{"管理员","宿管员","学生"}));
    Container c=getContentPane();
    c.setLayout(null);
    c.add(jbl_user_name);
    jbl_user_name.setBounds(20, 20, 80, 30);
    c.add(jtf_user_name);
    jtf_user_name.setBounds(100, 20, 200, 30);
    c.add(jbl_user_password);
    jbl_user_password.setBounds(20, 60, 80, 30);
    c.add(jpf_user_password);
    jpf_user_password.setBounds(100,60,200,30);
    c.add(jbl_user_role);
    jbl_user_role.setBounds(20, 100, 80, 30);
    c.add(jcb_user_role);
    jcb_user_role.setBounds(100, 100, 200, 30);
    c.add(jb_login);
    jb_login.setBounds(20, 140, 80, 30);
    c.add(jb_reset);
    jb_reset.setBounds(140, 140, 80, 30);
    c.add(jbl_error);
    jbl_error.setBounds(20, 180, 200, 30);
    jbl_error.setForeground(Color.RED);
    jb_login.addActionListener(this);
    jb_reset.addActionListener(this);
    this.setTitle("登录");
    this.setSize(350, 300);
    this.setLocationRelativeTo(null);
    this.setVisible(true);
    this.setDefaultCloseOperation(JFrame.EXIT_ON_CLOSE);
}
/*
 * 按钮事件处理
 */
@Override
public void actionPerformed(ActionEvent e) {
    String cmd=e.getActionCommand();
    if(cmd.equals("登录")){
        if(jtf_user_name.getText()==null||jtf_user_name.getText().equals("")){
            jbl_error.setText("请填写用户名!");
            return;
        }
```

```java
            UserBean user=new UserBean();
            user.setUser_name(jtf_user_name.getText());
            user.setUser_password(jpf_user_password.getText());
            UserDao userDao=new UserDao();
            userDao.chcekLogin(user);
            if(user.getErr_message()!=null){
                jbl_error.setText(user.getErr_message());
            }else{
                new MainFrame();
                this.dispose();
            }
        }else{
            jtf_user_name.setText("");
            jpf_user_password.setText("");
            jbl_error.setText("");
        }
    }
    /**
     * main:程序运行主函数.<br/>
     * @author      Administrator
     * @param args
     * @since JDK 1.6
     */
    public static void main(String[] args) {
        new LoginForm();
    }
}
```

MainFrame.java 代码

```java
package com.software.ui;
import java.awt.event.ActionEvent;
import java.awt.event.ActionListener;
import java.awt.event.MouseAdapter;
import java.awt.event.MouseEvent;
import javax.swing.JFrame;
import javax.swing.JMenu;
import javax.swing.JMenuBar;
import javax.swing.JMenuItem;
import javax.swing.event.MenuEvent;
import javax.swing.event.MenuListener;
/**
 * 类名：MainFrame<br/>
 * 功能：登录成功后显示带菜单的主窗体类.<br/>
```

```java
 * 创建时间：2016-6-1 上午11:29:51 <br/>
 * @author Administrator
 * @version
 * @since JDK 1.6
 */
public class MainFrame extends JFrame{      /**
     * 创建一个新的实例 MainFrame,对界面控件(菜单)进行布局.
     */
    public MainFrame(){
        JMenuBar mb = new JMenuBar();
        JMenu m_dorimtory = new JMenu("宿舍管理");
        JMenu m_student = new JMenu("学生管理");
        JMenu m_dorimtory_manager = new JMenu("宿管员管理");
        mb.add(m_dorimtory);
        mb.add(m_student);
        mb.add(m_dorimtory_manager);
        MyEvent me=new MyEvent();
        m_dorimtory.addMouseListener(me);
        m_student.addMouseListener(me);
        m_dorimtory_manager.addMouseListener(me);
        this.setJMenuBar(mb);
        this.setTitle("Main");
        this.setSize(800,600);
        this.setLocationRelativeTo(null);
        this.setVisible(true);
        this.setDefaultCloseOperation(JFrame.EXIT_ON_CLOSE);
    }
    /**
     * 菜单鼠标点击事件处理
     **/
    class MyEvent extends MouseAdapter{
        public void mouseClicked(MouseEvent e){
            String cmd=((JMenu)e.getSource()).getText();
            if(cmd.equals("宿舍管理")){
                new DormInfoFrame(MainFrame.this);
            }else if (cmd.equals("学生管理")){
            }else{
            }
        }
    }
}
```

项目十八

代码如下：
创建数据库代码

```sql
CREATE DATABASE DormDB CHARACTER SET gbk;            -- 创建数据库
USE DormDB；                -- 切换到当前数据库

CREATE TABLE t_student(
    student_id int NOT NULL PRIMARY KEY,      -- 学生编号
    student_name VARCHAR(20) NOT NULL,        -- 学生姓名
    room_id INT                               -- 寝室编号
);
ALTER TABLE t_student ADD CONSTRAINT FK_t_student    -- 创建外键
FOREIGN KEY (room_id) REFERENCES t_room(room_id);
INSERT INTO t_student
SELECT 1,'stu1',1 UNION
SELECT 2,'stu2',2 UNION
SELECT 3,'stu3',NULL;

CREATE TABLE t_user(
    user_id int NOT NULL PRIMARY KEY,         -- 用户编号
    user_name VARCHAR(20) NOT NULL,           -- 登录名
    user_password VARCHAR(12) NOT NULL,       -- 密码
    user_role char(1) NOT NULL                -- 角色
);
INSERT INTO t_user
SELECT 1,'admin','admin','1';

CREATE TABLE t_dormitory_building(
    dormitory_building_id INT NOT NULL PRIMARY KEY,       -- 宿舍楼编号
    dormitory_building_name VARCHAR(64) NOT NULL,         -- 宿舍楼名称
    bed_number int NOT NULL                               -- 本栋楼的寝室的床位数
);

INSERT INTO t_dormitory_building
SELECT 1,'1栋',10 UNION
SELECT 2,'2栋',6 ;

CREATE TABLE t_room(
    room_id   INT NOT NULL PRIMARY KEY,       -- 寝室编号
```

```sql
    room_name VARCHAR(64) NOT NULL,              -- 寝室名称
    living_number INT NOT NULL,                   -- 已住人数
    dormitory_building_id INT NOT NULL            -- 所属宿舍楼编号
);
ALTER TABLE t_room ADD CONSTRAINT FK_t_room
FOREIGN KEY(dormitory_building_id) REFERENCES t_dormitory_building(dormitory_building_id);
INSERT INTO t_room
SELECT 1,'1-101',3,1 UNION
SELECT 2,'2-202',6,2 UNION
SELECT 3,'2-203',2,2;
```

DBUtil.java 代码 同项目十七
MyTableModel.java 代码 同项目十七
UserDao.java 代码 同项目十七
UserBean.java 代码 同项目十七
LoginForm.java 代码 同项目十七

StudentInfoDao.java 代码

```java
package com.software.dao;
import java.sql.Connection;
import java.sql.ResultSet;
import java.sql.Statement;
import java.util.Vector;
import static com.software.dao.DBUtil.*;
/**
 * 类名：StudentInfoDao <br/>
 * 功能：封装学生住宿信息对数据库的操作类. <br/>
 * 创建时间：2016-6-1 下午 3:28:59 <br/>
 * @author Administrator
 * @version
 * @since JDK 1.6
 */
public class StudentInfoDao {
    /**
     * getColumnNames:获取显示学生住宿信息表格列标题. <br/>
     * @author Administrator
     * @return 学生住宿信息表格列标题
     * @since JDK 1.6
     */
    public Vector<String> getColumnNames(){
        Vector<String> columnNames=new Vector<String>();
        columnNames.add(0,"编号");
        columnNames.add(1,"姓名");
```

```java
        columnNames.add(2,"寝室名称");
        return columnNames;
}
/**
 * queryStudentInfo:查询学生住宿信息.<br/>
 * @author
Administrator
 * @param fullFlag 值为1查询未住满的寝室,其它值时查询所有寝室
 * @return 学生住宿信息返回结果
 * @since JDK 1.6
 */
public Vector queryStudentInfo(int fullFlag){
        Vector data = new Vector();
        Connection conn=null;
        Statement stm=null;
        ResultSet rs=null;
        String sql="select a.*,b.room_name from t_student as a LEFT JOIN t_room as b on a.room_id=b.room_id ";
        try{
                if(fullFlag==1){
                        sql=sql+" where a.room_id is null";
                }else{
                        sql=sql+" where b.room_id is not null";
                }
                conn=getConnection();
                stm=conn.createStatement();
                rs=stm.executeQuery(sql);
                while(rs.next()){
                        Vector<String> row = new Vector<String>();
                        row.add(rs.getString("student_id"));
                        row.add(rs.getString("student_name"));
                        String room_name=rs.getString("room_name");
                        if (room_name==null) room_name="";
                        row.add(room_name);
                        data.add(row);
                }
        } catch (Exception e) {
                System.out.println(sql);
                e.printStackTrace();
        }finally{
                try{
                        if(rs!=null) rs.close();
                        if(stm!=null) stm.close();
```

```java
                if(conn!=null) conn.close();
            }catch(Exception e){
                e.printStackTrace();
            }
        }
        return data;
    }
}
```

StudentInfoFrame.java 代码

```java
package com.software.ui;
import java.awt.BorderLayout;
import java.awt.Container;
import java.awt.event.ActionEvent;
import java.awt.event.ActionListener;
import java.awt.event.WindowAdapter;
import java.awt.event.WindowEvent;
import java.util.Vector;
import javax.swing.JButton;
import javax.swing.JFrame;
import javax.swing.JPanel;
import javax.swing.JScrollPane;
import javax.swing.JTable;
import com.software.dao.DormInfoDao;
import com.software.dao.MyTableModel;
import com.software.dao.StudentInfoDao;
/**
 * 类名：StudentInfoFrame <br/>
 * 功能：查询学生住宿情况窗体类. <br/>
 * 创建时间：2016-6-1 下午 3:38:18 <br/>
 *
 * @author Administrator
 * @version
 * @since JDK 1.6
 */
public class StudentInfoFrame extends JFrame implements ActionListener {
    JFrame parent;
    JPanel jp_top=new JPanel();
    JButton jb_notLive=new JButton("查询未分配的学生");
    JButton jb_live=new JButton("查询已分配的学生");
    JTable jtable=new JTable();
    /**
     * 创建一个新的实例 StudentInfoFrame,对界面控件进行布局.
```

```java
 * @param parent 打开弹出窗体上级窗体实例
 */
public StudentInfoFrame(JFrame parent){
    this.parent=parent;
    jp_top.add(jb_notLive);
    jp_top.add(jb_live);
    JScrollPane jsp=new JScrollPane();
    jsp.setViewportView(jtable);
    Container c=getContentPane();
    c.add(jp_top,BorderLayout.NORTH);
    c.add(jsp);
    jb_notLive.addActionListener(this);
    jb_live.addActionListener(this);
    this.setTitle("StudentInfoMain");
    this.setSize(400,300);
    this.setVisible(true);
    this.setLocationRelativeTo(parent);
    this.addWindowListener(new WindowAdapter(){
        public void windowClosing(WindowEvent e) {
            StudentInfoFrame.this.dispose();
        }
    });
    queryStudentInfo(0);
}
/**
 * queryStudentInfo:查询学生住宿信息.<br/>
 * @author       Administrator
 * @param liveFlag
 * @since JDK 1.6
 */
public void queryStudentInfo(int liveFlag){
    StudentInfoDao studentInfoDao=new StudentInfoDao();
    Vector data=studentInfoDao.queryStudentInfo(liveFlag);
    Vector columnNames=studentInfoDao.getColumnNames();
    MyTableModel tableModel=new MyTableModel(data,columnNames);
    jtable.setModel(tableModel);
}
/*
 * 按钮事件处理
 */
@Override
public void actionPerformed(ActionEvent e) {
    String cmd=e.getActionCommand();
```

```java
        if(cmd.equals("查询未分配的学生")){
            queryStudentInfo(1);
        }else if(cmd.equals("查询已分配的学生")){
            queryStudentInfo(0);
        }
    }
}
```

MainFrame.java 代码

```java
package com.software.ui;
import java.awt.event.ActionEvent;
import java.awt.event.ActionListener;
import java.awt.event.MouseAdapter;
import java.awt.event.MouseEvent;
import javax.swing.JFrame;
import javax.swing.JMenu;
import javax.swing.JMenuBar;
import javax.swing.JMenuItem;
import javax.swing.event.MenuEvent;
import javax.swing.event.MenuListener;
/**
 * 类名：MainFrame <br/>
 * 功能：登录成功后显示带菜单的主窗体类。<br/>
 * 创建时间：2016-6-1 上午 11:29:51 <br/>
 * @author Administrator
 * @version
 * @since JDK 1.6
 */
public class MainFrame extends JFrame{
    /**
     * 创建一个新的实例 MainFrame,对界面控件(菜单)进行布局.
     */
    public MainFrame(){
        JMenuBar mb = new JMenuBar();
        JMenu m_dorimtory = new JMenu("宿舍管理");
        JMenu m_student = new JMenu("学生管理");
        JMenu m_dorimtory_manager = new JMenu("宿管员管理");
        mb.add(m_dorimtory);
        mb.add(m_student);
        mb.add(m_dorimtory_manager);
        MyEvent me=new MyEvent();
        m_dorimtory.addMouseListener(me);
        m_student.addMouseListener(me);
```

```java
            m_dorimtory_manager.addMouseListener(me);
        this.setJMenuBar(mb);
        this.setTitle("Main");
        this.setSize(800,600);
        this.setLocationRelativeTo(null);
        this.setVisible(true);
        this.setDefaultCloseOperation(JFrame.EXIT_ON_CLOSE);
    }
    /**
     * 菜单鼠标点击事件处理
     **/
    class MyEvent extends MouseAdapter{
        public void mouseClicked(MouseEvent e){
            String cmd=((JMenu)e.getSource()).getText();
            if(cmd.equals("宿舍管理")){
                new DormInfoFrame(MainFrame.this);
            }else if (cmd.equals("学生管理")){
                new StudentInfoFrame(MainFrame.this);
            }else{
            }
        }
    }
}
```

项目十九

代码如下：
创建数据库代码

```
CREATE DATABASE DormDB CHARACTER         —— 创建数据库
SET gbk;
USE DormDB;     —— 切换到当前数据库

CREATE TABLE t_room_manager(
    room_manager_id INT NOT NULL PRIMARY KEY,    —— 宿管员编号
    room_manager_name VARCHAR(20) NOT NULL,      —— 宿管员姓名
    Dormitory_building_id INT                    —— 宿舍楼编号
);
ALTER TABLE t_room_manager ADD CONSTRAINT Fk_t_room_manager —— 添加外键
FOREIGN          KEY(Dormitory_building_id)
                                REFERENCES t_dormitory_building(dormitory_building_id);
INSERT INTO t_room_manager
SELECT 1,刘老师,1 UNION
```

```sql
SELECT 2,'王老师',2 UNION
SELECT 3,'李老师',NULL;

CREATE TABLE t_user(
    user_id int NOT NULL PRIMARY KEY,           -- 用户编号
    user_name VARCHAR(20) NOT NULL,             -- 登录名
    user_password VARCHAR(12) NOT NULL,         -- 密码
    user_role char(1) NOT NULL                  -- 角色
);
INSERT INTO t_user
SELECT 1,'admin','admin','1';

CREATE TABLE t_dormitory_building(
    dormitory_building_id INT NOT NULL PRIMARY KEY,    -- 宿舍楼编号
    dormitory_building_name VARCHAR(64) NOT NULL,      -- 宿舍楼名称
    bed_number int NOT NULL                            -- 本栋楼的寝室的床位数
);
INSERT INTO t_dormitory_building
SELECT 1,'1栋',10 UNION
SELECT 2,'2栋',6 UNION
SELECT 3,'3栋',8;
```

DBUtil.java 代码 同项目十七
MyTableModel.java 代码 同项目十七
UserDao.java 代码 同项目十七
UserBean.java 代码 同项目十七
LoginForm.java 代码 同项目十七

DormitoryBuildingDao.java 代码

```java
package com.software.dao;
import static com.software.dao.DBUtil.getConnection;
import java.sql.Connection;
import java.sql.ResultSet;
import java.sql.Statement;
import java.util.Vector;
import com.software.entity.DormitoryBuildingBean;
/**
 * 类名：DormitoryBuildingDao <br/>
 * 功能：封装对宿舍楼信息表数据操作类. <br/>
 * 创建时间：2016-6-1 下午 4:11:59 <br/>
 * @author Administrator
 * @version
 * @since JDK 1.6
```

```java
*/
public class DormitoryBuildingDao {
    /**
     * queryDormitoryBuilding:查询宿舍楼信息.<br/>
     * @author          Administrator
     * @return          宿舍楼信息查询结果
     * @since JDK 1.6
     */
    public Vector queryDormitoryBuilding(){
        Vector data =new Vector();
        Connection conn=null;
        Statement stm=null;
        ResultSet rs=null;
        String sql="SELECT * from t_dormitory_building";
        try {
            conn=getConnection();
            stm=conn.createStatement();
            rs=stm.executeQuery(sql);
            while(rs.next()){
                DormitoryBuildingBean dormitoryBuiling =new DormitoryBuildingBean();
                dormitoryBuiling.setBuildingID(rs.getInt("dormitory_building_id"));
                dormitoryBuiling.setBuidingName(rs.getString("dormitory_building_name"));
                dormitoryBuiling.setBedNumber(rs.getInt("bed_number"));
                data.add(dormitoryBuiling);
            }
        } catch (Exception e) {
            System.out.println(sql);
            e.printStackTrace();
        }finally{
            try{
                if(rs!=null) rs.close();
                if(stm!=null) stm.close();
                if(conn!=null) conn.close();
            }catch(Exception e){
                e.printStackTrace();
            }
        }
        return data;
    }
}
```

RoomManagerDao.java 代码

package com.software.dao;

```java
import static com.software.dao.DBUtil.getConnection;
import java.sql.Connection;
import java.sql.ResultSet;
import java.sql.Statement;
import java.util.Vector;
import java.sql.PreparedStatement;
import com.software.entity.DormitoryBuildingBean;
import com.software.entity.RoomManagerBean;
/**
 * 类名：RoomManagerDao <br/>
 * 功能：封装对宿管员信息表数据操作类. <br/>
 * 创建时间：2016-6-1 下午 4:16:53 <br/>
 * @author Administrator
 * @version
 * @since JDK 1.6
 */
public class RoomManagerDao {
    /**
     * queryRoomManager：查询宿管员信息. <br/>
     * @author       Administrator
     * @return 宿管员信息查询结果
     * @since JDK 1.6
     */
    public Vector queryRoomManager(){
        Vector data = new Vector();
        Connection conn = null;
        Statement stm = null;
        ResultSet rs = null;
        String sql = "SELECT * from t_room_manager";
        try {
            conn = getConnection();
            stm = conn.createStatement();
            rs = stm.executeQuery(sql);
            while(rs.next()){
                RoomManagerBean roomManager = new RoomManagerBean();
                roomManager.setManagerId(rs.getInt("room_manager_id"));
                roomManager.setManagerName(rs.getString("room_manager_name"));
                roomManager.setBuildingID(rs.getInt("Dormitory_building_id"));
                data.add(roomManager);
            }
        } catch (Exception e) {
            System.out.println(sql);
            e.printStackTrace();
```

```java
        }finally{
            try{
                if(rs!=null) rs.close();
                if(stm!=null) stm.close();
                if(conn!=null) conn.close();
            }catch(Exception e){
                e.printStackTrace();
            }
        }
        return data;
    }
    /**
     * updateRoomManager:修改宿管员信息.<br/>
     * @author    Administrator
     * @param roomManager 宿管员实体对象
     * @return 是否修改成功
     * @since JDK 1.6
     */
    public boolean updateRoomManager(RoomManagerBean roomManager){
        boolean isSuccess=false;
        Connection conn=null;
        PreparedStatement stm=null;
        String sql="update t_room_manager set Dormitory_building_id=? where room_manager_id=?";
        try {
            conn=getConnection();
            stm=conn.prepareStatement(sql);
            stm.setInt(1, roomManager.getBuildingID());
            stm.setInt(2, roomManager.getManagerId());
            if(stm.executeUpdate()>0) isSuccess=true;
        } catch (Exception e) {
            System.out.println(sql);
            e.printStackTrace();
        }finally{
            try{
                if(stm!=null) stm.close();
                if(conn!=null) conn.close();
            }catch(Exception e){
                e.printStackTrace();
            }
        }
        return isSuccess;
    }
}
```

DormitoryBuildingBean.java 代码

```java
package com.software.entity;
/**
 * 类名：DormitoryBuildingBean <br/>
 * 功能：对应数据库中宿舍楼信息表(t_dormitory_building)实体类.<br/>
 * 创建时间：2016-6-1 下午4:24:33 <br/>
 * @author Administrator
 * @version
 * @since JDK 1.6
 */
public class DormitoryBuildingBean {
    private int buildingID;
    private String buidingName;
    private int bedNumber;
    public int getBuildingID() {
        return buildingID;
    }
    public void setBuildingID(int buildingID) {
        this.buildingID = buildingID;
    }
    public String getBuidingName() {
        return buidingName;
    }
    public void setBuidingName(String buidingName) {
        this.buidingName = buidingName;
    }
    public int getBedNumber() {
        return bedNumber;
    }
    public void setBedNumber(int bedNumber) {
        this.bedNumber = bedNumber;
    }
    /*
     * 重写 toString 方法用于下拉列表
     */
    @Override
    public String toString() {
        return  buidingName;
    }
    /*
     * 重写 equals 方法用于下拉列表
     */
```

```java
@Override
public boolean equals(Object obj) {
    DormitoryBuildingBean dormitoryBuiding=(DormitoryBuildingBean)obj;
    return dormitoryBuiding.getBuildingID()==buildingID;
}
}
```

RoomManagerBean.java 代码

```java
package com.software.entity;
/**
 * 类名：RoomManagerBean <br/>
 * 功能：对应数据库中宿管员信息表(t_room_manager)实体类。<br/>
 * 创建时间：2016-6-1 下午4:28:04 <br/>
 * @author Administrator
 * @version
 * @since JDK 1.6
 */
public class RoomManagerBean {
    private int managerId;
    private String managerName;
    private Integer buildingID;
    public int getManagerId() {
        return managerId;
    }
    public void setManagerId(int managerId) {
        this.managerId = managerId;
    }
    public String getManagerName() {
        return managerName;
    }
    public void setManagerName(String managerName) {
        this.managerName = managerName;
    }
    public Integer getBuildingID() {
        return buildingID;
    }
    public void setBuildingID(Integer buildingID) {
        this.buildingID = buildingID;
    }
    /*
     * 重写toString方法用于下拉列表
     */
    @Override
```

```java
    public String toString() {
        return managerName;
    }
    /*
     * 重写equals方法用于下拉列表
     */
    @Override
    public boolean equals(Object obj) {
        RoomManagerBean roomManager=(RoomManagerBean)obj;
        return  roomManager.getBuildingID()==buildingID;
    }

}
```

RoomManagerMain.java 代码

```java
package com.software.ui;
import java.awt.Container;
import java.awt.GridLayout;
import java.awt.event.ActionEvent;
import java.awt.event.ActionListener;
import java.awt.event.WindowAdapter;
import java.awt.event.WindowEvent;
import javax.swing.JButton;
import javax.swing.JComboBox;
import javax.swing.JFrame;
import javax.swing.JLabel;
import javax.swing.JOptionPane;
import javax.swing.JPanel;
import com.software.dao.DormitoryBuildingDao;
import com.software.dao.RoomManagerDao;
import com.software.entity.DormitoryBuildingBean;
import com.software.entity.RoomManagerBean;
/**
 * 类名:RoomManagerMain <br/>
 * 功能:宿管员管理窗体类.<br/>
 * 创建时间:2016-6-1 下午4:31:38 <br/>
 * @author Administrator
 * @version
 * @since JDK 1.6
 */
public class RoomManagerMain extends JFrame implements ActionListener{
    JPanel jp_jcb=new JPanel();
    JPanel jp_button=new JPanel();
```

```java
JLabel jbl_DormitoryBuild=new JLabel("选择宿舍楼");
JComboBox jcb_DormitoryBuild;
JLabel jbl_RoomManager=new JLabel("选择宿管员");
JComboBox jcb_RoomManager;
JButton jb_assign=new JButton("分配");
DormitoryBuildingDao dormitoryBuildingDao=new DormitoryBuildingDao();
RoomManagerDao roomManagerDao=new RoomManagerDao();
/**
 * 创建一个新的实例 RoomManagerMain,对界面控件进行布局.
 * @param parent 弹出窗体的上一级窗体实例
 */
public RoomManagerMain(JFrame parent){
    jcb_DormitoryBuild=new JComboBox(dormitoryBuildingDao.queryDormitoryBuilding());
    jcb_RoomManager=new JComboBox(roomManagerDao.queryRoomManager());
    jp_jcb.add(jbl_DormitoryBuild);
    jp_jcb.add(jcb_DormitoryBuild);
    jp_jcb.add(jbl_RoomManager);
    jp_jcb.add(jcb_RoomManager);
    jp_button.add(jb_assign);
    jb_assign.addActionListener(this);
    Container c=getContentPane();
    c.setLayout(new GridLayout(2,1));
    c.add(jp_jcb);
    c.add(jp_button);
    this.setTitle("RoomManagerMain");
    this.setSize(400,300);
    this.setVisible(true);
    this.setLocationRelativeTo(parent);
    this.addWindowListener(new WindowAdapter(){
        public void windowClosing(WindowEvent e) {
            RoomManagerMain.this.dispose();
        }
    });
}
/*
 * 按钮事件处理
 */
@Override
public void actionPerformed(ActionEvent e) {
    DormitoryBuildingBean dormitoryBuild=(DormitoryBuildingBean)jcb_DormitoryBuild.getSelectedItem();
    RoomManagerBean roomManager=(RoomManagerBean)jcb_RoomManager.getSelectedItem();
    roomManager.setBuildingID(dormitoryBuild.getBuildingID());
```

```java
        if(roomManagerDao.updateRoomManager(roomManager))
            JOptionPane.showMessageDialog(this,"分配成功");
        else
            JOptionPane.showMessageDialog(this,"分配失败");
    }

}
```

MainFrame.java 代码

```java
package com.software.ui;
import java.awt.event.ActionEvent;
import java.awt.event.ActionListener;
import java.awt.event.MouseAdapter;
import java.awt.event.MouseEvent;
import javax.swing.JFrame;
import javax.swing.JMenu;
import javax.swing.JMenuBar;
import javax.swing.JMenuItem;
import javax.swing.event.MenuEvent;
import javax.swing.event.MenuListener;
/**
 * 类名：MainFrame <br/>
 * 功能：登录成功后显示带菜单的主窗体类.<br/>
 * 创建时间：2016-6-1 上午 11:29:51 <br/>
 * @author Administrator
 * @version
 * @since JDK 1.6
 */
public class MainFrame extends JFrame{
    /**
     * 创建一个新的实例 MainFrame,对界面控件(菜单)进行布局.
     */
    public MainFrame(){
        JMenuBar mb = new JMenuBar();
        JMenu m_dorimtory = new JMenu("宿舍管理");
        JMenu m_student = new JMenu("学生管理");
        JMenu m_dorimtory_manager = new JMenu("宿管员管理");
        mb.add(m_dorimtory);
        mb.add(m_student);
        mb.add(m_dorimtory_manager);
        MyEvent me=new MyEvent();
        m_dorimtory.addMouseListener(me);
        m_student.addMouseListener(me);
```

```java
            m_dorimtory_manager.addMouseListener(me);
            this.setJMenuBar(mb);
            this.setTitle("Main");
            this.setSize(800,600);
            this.setLocationRelativeTo(null);
            this.setVisible(true);
            this.setDefaultCloseOperation(JFrame.EXIT_ON_CLOSE);
        }
        /**
         * 菜单鼠标点击事件处理
         */
        class MyEvent extends MouseAdapter{
            public void mouseClicked(MouseEvent e){
                String cmd=((JMenu)e.getSource()).getText();
                if(cmd.equals("宿舍管理")){
                    new DormInfoFrame(MainFrame.this);
                }else if (cmd.equals("学生管理")){
                    new StudentInfoFrame(MainFrame.this);
                }else{
                    new RoomManagerMain(MainFrame.this);
                }
            }
        }
}
```

项目二十

代码如下：
创建数据库代码
```
CREATE DATABASE oadb CHARACTER SET GBK;
USE oadb;

CREATE TABLE t_category(
    category_id VARCHAR(20) NOT NULL PRIMARY KEY,    -- 办公用品类别编号
    category_name VARCHAR(20) NOT NULL               -- 办公用品类别名称
);
INSERT INTO t_category
SELECT '001','文具' UNION
SELECT '002','耗材' UNION
SELECT '003','纸张';

CREATE TABLE t_product(
```

```
    product_id VARCHAR(20) NOT NULL PRIMARY KEY,    -- 办公用品编号
    product_name VARCHAR(20) NOT NULL,              -- 名称
    category_id VARCHAR(20) NOT NULL,               -- 类别
    product_number FLOAT NOT NULL,                  -- 数量
    product_price FLOAT NOT NULL                    -- 单价
);
INSERT INTO t_product
SELECT '001','签字笔','001',2,1.0 UNION
SELECT '002','2B 铅笔','001',4,1.5 UNION
SELECT '003','订书机','001',5,25.0;
```

DBUtil.java 代码

```java
package com.software.dao;
import java.sql.Connection;
import java.sql.DriverManager;
import java.sql.SQLException;
/**
 * 类名：DBUtil <br/>
 * 功能：自定义数据库工具类,封装通用数据库操作 <br/>
 * 创建时间：2016-5-31 下午 4:51:22 <br/>
 * @author Administrator
 * @version
 * @since JDK 1.6
 */
public class DBUtil {
    /**
     * getConnection:连接数据库返回,数据库连接对象. <br/>
     * @author     Administrator
     * @return 数据库连接对象
     * @throws ClassNotFoundException
     * @throws SQLException
     * @since JDK 1.6
     */
    public static Connection getConnection() throws ClassNotFoundException,SQLException{
        String url="jdbc:mysql://127.0.0.1:8306/oadb";
        String user="root";
        String pwd="1234";
        Class.forName("com.mysql.jdbc.Driver");
        Connection conn=DriverManager.getConnection(url, user, pwd);
        return conn;
    }

}
```

MyTableModel.java 代码

```java
package com.software.dao;
import java.util.Vector;
import javax.swing.table.DefaultTableModel;
/**
 * 类名：MyTableModel <br/>
 * 功能：自定义JTable表格数据模型代码. <br/>
 * 创建时间：2016-5-31 下午5:03:10 <br/>
 * @author Administrator
 * @version
 * @since JDK 1.6
 */
public class MyTableModel extends DefaultTableModel {
    /**
     * 创建一个新的实例 MyTableModel.
     * @param data 表格数据
     * @param columnNames 显示表格的列名
     */
    public MyTableModel(Vector data, Vector columnNames) {
        super(data, columnNames);
    }
    /**
     * 控制表格单元格是否可编辑.
     * @param r 行号
     * @param c 列号
     */
    public boolean isCellEditable(int r, int c) {
        return false;
    }
    /**
     * 获得单元格列类型.
     * @param c 列号
     * @return 列类型
     */
    public Class getColumnClass(int c) {
        return getValueAt(0, c).getClass();
    }
}
```

CategoryDao.java 代码

```java
package com.software.dao;
import static com.software.dao.DBUtil.getConnection;
```

```java
import java.sql.Connection;
import java.sql.PreparedStatement;
import java.sql.SQLException;
import com.software.entity.CategoryBean;
/**
 * 类名：CategoryDao <br/>
 * 功能：封装对办公用品类别表的数据库操作类。<br/>
 * 创建时间：2016-6-1 下午8:59:31 <br/>
 * @author Administrator
 * @version
 * @since JDK 1.6
 */
public class CategoryDao {
    /**
     * insertCategory:添加办公用品类别。<br/>
     * @author      Administrator
     * @param category 办公用品类别实体类
     * @return 添加是否成功
     * @since JDK 1.6
     */
    public boolean insertCategory(CategoryBean category) {
        boolean isSuccess = false;
        Connection conn = null;
        PreparedStatement ps = null;
        try {
            String sql = "insert into t_category values(?,?)";
            conn = getConnection();
            ps = conn.prepareStatement(sql);
            ps.setString(1, category.getCategoryID());
            ps.setString(2, category.getCategoryName());
            ps.executeUpdate();
            isSuccess = true;
        } catch (ClassNotFoundException e) {
            e.printStackTrace();
        } catch (SQLException e) {
            e.printStackTrace();
        } finally {
            try {
                if (ps != null)
                    ps.close();
                if (conn != null)
                    conn.close();
            } catch (SQLException e) {
```

```
                    e.printStackTrace();
            }
        }
        return isSuccess;
    }
}
```

ProductDao.java 代码

```
package com.software.dao;
import static com.software.dao.DBUtil.getConnection;
import java.sql.Connection;
import java.sql.ResultSet;
import java.sql.SQLException;
import java.sql.Statement;
import java.util.Vector;
import javax.swing.table.DefaultTableModel;
import javax.swing.table.TableModel;
import com.software.entity.CategoryBean;
import com.software.entity.ProductBean;
/**
 * 类名：ProductDao <br/>
 * 功能：封装对办公用品信息表的数据库操作类。<br/>
 * 创建时间：2016-6-1 下午 9:10:39 <br/>
 * @author Administrator
 * @version
 * @since JDK 1.6
 */
public class ProductDao {
    /**
     * getColumnNames：获取显示办公用品表格中文列标题。<br/>
     * @author      Administrator
     * @return 办公用品表格中文列标题
     * @since JDK 1.6
     */
    public static Vector<String> getColumnNames(){
        Vector<String> columnNames=new Vector<String>();
        columnNames.add(0,"编号");
        columnNames.add(1,"名称");
        return columnNames;
    }
    /**
     * getEmpty：获取一个带列标题办公用品空表数据模型。<br/>
     * @author      Administrator
```

```java
 * @return 一个带列标题办公用品空表数据模型
 * @since JDK 1.6
 */
public static TableModel getEmpty(){
    Vector data=new Vector();
    DefaultTableModel dmt=new DefaultTableModel(data,getColumnNames());
    return dmt;
}
/**
 * queryProduct:依据办公用品编号或名称查询办公用品信息.<br/>
 * @author        Administrator
 * @param productID 办公用品编号
 * @param productName 办公用品名称
 * @return 办公用品信息查询结果
 * @since JDK 1.6
 */
public Vector queryProduct(String productID,String productName){
    Vector data=new Vector();
    Connection conn=null;
    Statement stm=null;
    ResultSet rs=null;
    String sql="select * from t_product where (1=1) ";
    try {
        if(!(productID==null||productID.equals(""))){
            if(!(productName==null||productName.equals(""))){
                sql+=" and (product_id='"+productID+"' or product_name='"+productName+"')";
            }else{
                sql+=" and (product_id='"+productID+"')";
            }
        }else if(!(productName==null||productName.equals(""))){
            {
                sql+=" and (product_name='"+productName+"')";
            }
        }
        conn=getConnection();
        stm=conn.createStatement();
        rs=stm.executeQuery(sql);
        while(rs.next()){
            Vector row=new Vector();
            row.add(rs.getString("product_id"));
            row.add(rs.getString("product_name"));
            data.add(row);
```

```java
                }
            } catch (ClassNotFoundException e) {
                e.printStackTrace();
            } catch (SQLException e) {
                System.out.println(sql);
                e.printStackTrace();
            } finally{
                try {
                    if(rs! = null) rs.close();
                    if(stm! = null) stm.close();
                    if(conn! = null) conn.close();
                } catch (SQLException e) {
                    e.printStackTrace();
                }
            }
            return data;
        }
        /**
         * querySingleProduct:依据办公用品编号查询办公用品信息. <br/>
         * @author         Administrator
         * @param productId 办公用品编号
         * @return 办公用品信息查询结果
         * @since JDK 1.6
         */
        public ProductBean querySingleProduct(String productId){
            ProductBean product = new ProductBean();
            Connection conn = null;
            Statement stm = null;
            ResultSet rs = null;
            String sql = "select a.* ,b.category_name from t_product as a,t_category as b where a.category_id= b.category_id"
                    +" and  a.product_id='"+productId+"'";
            try {
                conn = getConnection();
                stm = conn.createStatement();
                rs = stm.executeQuery(sql);
                while(rs.next()){
                    product.setProductID(rs.getString("product_id"));
                    product.setProductName(rs.getString("product_name"));
                    product.setProductNumber(rs.getFloat("product_number"));
                    product.setProductPrice(rs.getFloat("product_price"));
                    CategoryBean category = new CategoryBean(rs.getString("category_id"), rs.getString("category_name"));
```

```java
                product.setCategory(category);
            }
        } catch (ClassNotFoundException e) {
            e.printStackTrace();
        } catch (SQLException e) {
            System.out.println(sql);
            e.printStackTrace();
        }finally{
            try {
                if(rs!=null) rs.close();
                if(stm!=null) stm.close();
                if(conn!=null) conn.close();
            } catch (SQLException e) {
                e.printStackTrace();
            }
        }
        return product;
    }
}
```

CategoryBean.java 代码

```java
package com.software.entity;
/**
 * 类名：CategoryBean <br/>
 * 功能：对应办公用品分类表(t_category)的实体类。<br/>
 * 创建时间：2016-6-1 下午 9:12:04 <br/>
 * @author Administrator
 * @version
 * @since JDK 1.6
 */
public class CategoryBean {
    private String categoryID;
    private String categoryName;
    public CategoryBean() {
    }
    public CategoryBean(String categoryID, String categoryName) {
        this.categoryID = categoryID;
        this.categoryName = categoryName;
    }
    public String getCategoryID() {
        return categoryID;
    }
    public void setCategoryID(String categoryID) {
```

```java
        this.categoryID = categoryID;
    }
    public String getCategoryName() {
        return categoryName;
    }
    public void setCategoryName(String categoryName) {
        this.categoryName = categoryName;
    }

}
```

ProductBean.java 代码

```java
package com.software.entity;
/**
 * 类名：ProductBean <br/>
 * 功能：对应办公用品信息表(t_product)实体类. <br/>
 * 创建时间：2016-6-1 下午9:13:43 <br/>
 * @author Administrator
 * @version
 * @since JDK 1.6
 */
public class ProductBean {
    private String productID;
    private String productName;
    private CategoryBean category;
    private float productNumber;
    private float productPrice;
    public String getProductID() {
        return productID;
    }
    public void setProductID(String productID) {
        this.productID = productID;
    }
    public String getProductName() {
        return productName;
    }
    public void setProductName(String productName) {
        this.productName = productName;
    }
    public CategoryBean getCategory() {
        return category;
    }
    public void setCategory(CategoryBean category) {
```

```java
        this.category = category;
    }
    public float getProductNumber() {
        return productNumber;
    }
    public void setProductNumber(float productNumber) {
        this.productNumber = productNumber;
    }
    public float getProductPrice() {
        return productPrice;
    }
    public void setProductPrice(float productPrice) {
        this.productPrice = productPrice;
    }

}
```

AddCategoryFrame.java 代码

```java
package com.software.ui;
import java.awt.Container;
import java.awt.event.ActionEvent;
import java.awt.event.ActionListener;
import java.awt.event.WindowAdapter;
import java.awt.event.WindowEvent;
import javax.swing.JButton;
import javax.swing.JDialog;
import javax.swing.JFrame;
import javax.swing.JLabel;
import javax.swing.JOptionPane;
import javax.swing.JTextField;
import com.software.dao.CategoryDao;
import com.software.entity.CategoryBean;
/**
 * 类名：AddCategoryFrame <br/>
 * 功能：添加办公用品类别窗体类. <br/>
 * 创建时间：2016-6-1 下午 9:15:40 <br/>
 * @author Administrator
 * @version
 * @since JDK 1.6
 */
public class AddCategoryFrame extends JDialog implements ActionListener{
    JLabel jbl_id=new JLabel("类别编码：");
    JTextField jtf_id=new JTextField(30);
```

```java
JLabel jbl_name=new JLabel("类别名称:");
JTextField jtf_name=new JTextField(30);
JButton jb_save=new JButton("保存");
/**
 * 创建一个新的实例 AddCategoryFrame,为界面控件布局.
 * @param parent 打开弹出窗体上一级窗体的实例
 */
public  AddCategoryFrame(JFrame parent){
    super(parent,"办公用品类别",true);
    Container container=getContentPane();
    container.setLayout(null);
    container.add(jbl_id);
    jbl_id.setBounds(50,50,80,30);
    container.add(jtf_id);
    jtf_id.setBounds(150,50,200,30);
    container.add(jbl_name);
    jbl_name.setBounds(50,100,80,30);
    container.add(jtf_name);
    jtf_name.setBounds(150,100,200,30);
    container.add(jb_save);
    jb_save.setBounds(150,150,80,30);
    jb_save.addActionListener(this);
    this.setSize(400,300);
    this.setLocationRelativeTo(parent);
    this.setVisible(true);
    this.addWindowListener(new WindowAdapter() {
        @Override
        public void windowClosing(WindowEvent e) {
            dispose();
        }
    });
}
/**
 * saveCategory:保存办公用品类别信息. <br/>
 * @author         Administrator
 * @return 保存是否成功
 * @since JDK 1.6
 */
public boolean saveCategory(){
    CategoryBean category=new CategoryBean();
    CategoryDao categoryDao=new CategoryDao();
    category.setCategoryID(jtf_id.getText());
    category.setCategoryName(jtf_name.getText());
```

```java
        return categoryDao.insertCategory(category);
    }
    /*
     * 按钮事件处理
     */
    @Override
    public void actionPerformed(ActionEvent e) {
        if(saveCategory()){
            JOptionPane.showMessageDialog(null,"保存成功");
            this.dispose();
        }
        else{
            JOptionPane.showMessageDialog(null,"保存不成功请检查数据");
        }
    }
}
```

ProductDetailFrame.java 代码

```java
package com.software.ui;
import java.awt.Container;
import java.awt.event.ActionEvent;
import java.awt.event.ActionListener;
import java.awt.event.WindowAdapter;
import java.awt.event.WindowEvent;
import javax.swing.JButton;
import javax.swing.JDialog;
import javax.swing.JFrame;
import javax.swing.JLabel;
import import com.software.dao.ProductDao;
import com.software.entity.ProductBean;
/**
 * 类名：ProductDetailFrame <br/>
 * 功能：显示办公用品详细信息窗体类. <br/>
 * 创建时间：2016-6-1 下午 9:27:29 <br/>
 * @author Administrator
 * @version
 * @since JDK 1.6
 */
public class ProductDetailFrame extends JDialog implements ActionListener {
    /**
     * 创建一个新的实例 ProductDetailFrame,为界面控件布局.
     * @param parent 打开弹出窗体的上一级窗体
     * @param productID 办公用品编号
```

```java
 */
public ProductDetailFrame(JDialog parent,String productID){
    super(parent,"办公用品详细信息",true);
    ProductDao productDao=new ProductDao();
    ProductBean product=productDao.querySingleProduct(productID);
    JLabel jbl_id=new JLabel("办公用品编号   "+product.getProductID());
    JLabel jbl_name=new JLabel("名称        "+product.getProductName());
    JLabel jbl_category=new JLabel("类别      "+product.getCategory().getCategoryName());
    JLabel jbl_number=new JLabel("数量       "+product.getProductNumber());
    JLabel jbl_price=new JLabel("单价       "+product.getProductPrice());
    JButton jb_close=new JButton("关闭");
    Container container=getContentPane();
    container.setLayout(null);
    container.add(jbl_id);
    jbl_id.setBounds(20,20,200,30);
    container.add(jbl_name);
    jbl_name.setBounds(20,60,200,30);
    container.add(jbl_category);
    jbl_category.setBounds(20,100,200,30);
    container.add(jbl_number);
    jbl_number.setBounds(20,140,200,30);
    container.add(jbl_price);
    jbl_price.setBounds(20,180,200,30);
    container.add(jb_close);
    jb_close.setBounds(120,220,80,30);
    jb_close.addActionListener(this);
    this.setLocationRelativeTo(parent);
    this.setSize(300,350);
    this.setVisible(true);
    this.addWindowListener(new WindowAdapter() {
        @Override
        public void windowClosing(WindowEvent e) {
            dispose();
        }
    });
}
/*
 * 按钮事件处理
 */
@Override
public void actionPerformed(ActionEvent e) {
    this.dispose();
}
```

}

ProductQueryFrame.java 代码

```java
package com.software.ui;
import java.awt.BorderLayout;
import java.awt.Container;
import java.awt.Dimension;
import java.awt.event.ActionEvent;
import java.awt.event.ActionListener;
import java.awt.event.WindowAdapter;
import java.awt.event.WindowEvent;
import java.util.Vector;
import javax.swing.BorderFactory;
import javax.swing.JButton;
import javax.swing.JDialog;
import javax.swing.JFrame;
import javax.swing.JLabel;
import javax.swing.JPanel;
import javax.swing.JScrollPane;
import javax.swing.JTable;
import javax.swing.JTextField;
import javax.swing.ListSelectionModel;
import javax.swing.event.ListSelectionEvent;
import javax.swing.event.ListSelectionListener;
import com.software.dao.MyTableModel;
import com.software.dao.ProductDao;
/**
 * 类名：ProductQueryFrame <br/>
 * 功能：办公用品信息查询窗体类. <br/>
 * 创建时间：2016-6-1 下午 9:32:41 <br/>
 * @author Administrator
 * @version
 * @since JDK 1.6
 */
public class ProductQueryFrame extends JDialog implements ActionListener {
    JPanel jp_query = new JPanel();
    JLabel jbl_id = new JLabel("办公用品编号");
    JTextField jtf_id = new JTextField(30);
    JLabel jbl_name = new JLabel("办公用品名称");
    JTextField jtf_name = new JTextField(30);
    JButton jb_query = new JButton("查询");
    JTable table = null;
    /**
```

```java
 * 创建一个新的实例 ProductQueryFrame,为界面控件布局.
 * @param parent 打开弹出窗体上一级窗体的实例
 */
public ProductQueryFrame(JFrame parent) {
    super(parent, "办公用品查询", true);
    jp_query.setLayout(null);
    jp_query.add(jbl_id);
    jp_query.add(jtf_id);
    jbl_id.setBounds(20, 20, 80, 30);
    jtf_id.setBounds(100, 20, 200, 30);
    jp_query.add(jbl_name);
    jp_query.add(jtf_name);
    jbl_name.setBounds(20, 60, 80, 30);
    jtf_name.setBounds(100, 60, 200, 30);
    jp_query.add(jb_query);
    jb_query.setBounds(220, 100, 80, 30);
    jb_query.addActionListener(this);
    table = new JTable(ProductDao.getEmpty());
    table.getSelectionModel().setSelectionMode(ListSelectionModel.SINGLE_SELECTION);
    table.getSelectionModel().addListSelectionListener(new SelectionListener());
    jp_query.setPreferredSize(new Dimension(350, 200));
    JScrollPane jsp = new JScrollPane(table);
    jsp.setBorder(BorderFactory.createTitledBorder("办公用品信息"));
    Container container = getContentPane();
    container.add(jp_query, BorderLayout.NORTH);
    container.add(jsp);
    this.setSize(350, 500);
    this.setLocationRelativeTo(parent);
    this.setVisible(true);
    this.addWindowListener(new WindowAdapter() {
        @Override
        public void windowClosing(WindowEvent e) {
            dispose();
        }
    });
}
/*
 * 按钮事件的处理
 */
@Override
public void actionPerformed(ActionEvent e) {
    ProductDao productDao = new ProductDao();
    Vector data = productDao.queryProduct(jtf_id.getText(), jtf_name.getText());
```

```java
            MyTableModel myTableModel = new MyTableModel(data, ProductDao.getColumnNames());
            table.setModel(myTableModel);
    }
    /**
     * 类名:SelectionListener <br/>
     * 功能:表格行选择事件监听类. <br/>
     * 创建时间:2016-6-1 下午 9:33:00 <br/>
     * @author Administrator
     * @version ProductQueryFrame
     * @since JDK 1.6
     */
    class SelectionListener implements ListSelectionListener {
        public void valueChanged(ListSelectionEvent e) {
            int row = table.getSelectedRow();
            if(!e.getValueIsAdjusting()&&row>=0){
                new ProductDetailFrame(ProductQueryFrame.this,(String)table.getValueAt(row, 0));
                table.repaint();
            }
        }
    }
}
```

MainFrame.java 代码

```java
package com.software.ui;
import java.awt.Container;
import java.awt.event.ActionEvent;
import java.awt.event.ActionListener;
import javax.swing.JButton;
import javax.swing.JFrame;
/**
 * 类名:MainFrame <br/>
 * 功能:程序运行主窗体类. <br/>
 * 创建时间:2016-6-1 下午 9:23:26 <br/>
 * @author Administrator
 * @version
 * @since JDK 1.6
 */
public class MainFrame extends JFrame implements ActionListener{
    JButton jb_add=new JButton("类别添加");
    JButton jb_query=new JButton("办公用品查询");
    /**
```

* 创建一个新的实例 MainFrame,为界面控件布局.
 */
```java
public MainFrame(){
    Container c=getContentPane();
    c.setLayout(null);
    c.add(jb_add);
    jb_add.setBounds(50,100,200,30);
    c.add(jb_query);
    jb_query.setBounds(50,150,200,30);
    jb_add.addActionListener(this);
    jb_query.addActionListener(this);
    this.setSize(300,300);
    this.setTitle("通达办公自动化系统");
    this.setLocationRelativeTo(null);
    this.setVisible(true);
    this.setDefaultCloseOperation(JFrame.EXIT_ON_CLOSE);
}
/**
 * main:程序运行主函数.<br/>
 * @author    Administrator
 * @param args
 * @since JDK 1.6
 */
public static void main(String[] args) {
    new MainFrame();
}
/*
 * 按钮事件处理
 */
@Override
public void actionPerformed(ActionEvent e) {
    String cmd=e.getActionCommand();
    if(cmd.equals("类别添加")){
        new AddCategoryFrame(this);
    }else{
        new ProductQueryFrame(this);
    }
}
}
```

项目二十一

代码如下:

创建数据库代码
CREATE DATABASE oadb CHARACTER SET GBK;
USE oadb;

CREATE TABLE t_meeting_room(
    meeting_room_id VARCHAR(20) NOT NULL PRIMARY KEY, —— 会议室编号
    meeting_room_name VARCHAR(20) NOT NULL   —— 会议室名称
);
INSERT INTO t_meeting_room
SELECT '001','多媒体会议室' UNION
SELECT '002','多功能厅' UNION
SELECT '003','第三会议室';

CREATE TABLE t_reservation(
    reservation_id VARCHAR(20) NOT NULL PRIMARY KEY,   —— 会议室预订编号
    reservation_name VARCHAR(20) NOT NULL,   —— 预订人
    start_time datetime NOT NULL,   —— 开始时间
    end_time datetime NOT NULL,   —— 结束时间
    meeting_room_id VARCHAR(20) NOT NULL   —— 会议室
);
INSERT INTO t_reservation
SELECT '001','王明','2011-06-03 15:30','2011-06-03 17:50','001' UNION
SELECT '002','周文','2011-06-04 08:30','2011-06-04 09:50','001' UNION
SELECT '003','刘伟','2011-06-05 10:00','2011-06-05 12:00','001'

DBUtil.java 代码
package com.software.dao;
import java.sql.Connection;
import java.sql.DriverManager;
import java.sql.SQLException;
/**
 * 类名：DBUtil
 * 功能：自定义数据库工具类,封装通用数据库操作
 * 创建时间：2016-5-31 下午 4:51:22
 * @author Administrator
 * @version
 * @since JDK 1.6
 */
public class DBUtil {
    /**
     * getConnection:连接数据库返回,数据库连接对象.
     * @author   Administrator
     * @return 数据库连接对象

```java
 *  @throws ClassNotFoundException
 *  @throws SQLException
 *  @since JDK 1.6
 */
public static Connection getConnection() throws ClassNotFoundException,SQLException{
    String url="jdbc:mysql://127.0.0.1:8306/oadb";
    String user="root";
    String pwd="1234";
    Class.forName("com.mysql.jdbc.Driver");
    Connection conn=DriverManager.getConnection(url, user, pwd);
    return conn;
}
}
```

MyTableModel.java 代码

```java
package com.software.dao;
import java.util.Vector;
import javax.swing.table.DefaultTableModel;
/**
 * 类名：MyTableModel <br/>
 * 功能：自定义 JTable 表格数据模型代码. <br/>
 * 创建时间：2016-5-31 下午 5:03:10 <br/>
 * @author Administrator
 * @version
 * @since JDK 1.6
 */
public class MyTableModel extends DefaultTableModel {
    /**
     * 创建一个新的实例 MyTableModel.
     * @param data 表格数据
     * @param columnNames 显示表格的列名
     */
    public MyTableModel(Vector data, Vector columnNames) {
        super(data, columnNames);
    }
    /**
     * 控制表格单元格是否可编辑.
     * @param r 行号
     * @param c 列号
     */
    public boolean isCellEditable(int r, int c) {
        return false;
```

```java
    }
    /**
     * 获得单元格列类型.         * @param c 列号
     * @return 列类型
     */
    public Class getColumnClass(int c) {
        return getValueAt(0, c).getClass();
    }
}
```

MeetingRoomDao.java 代码

```java
package com.software.dao;
import static com.software.dao.DBUtil.getConnection;
import java.sql.Connection;
import java.sql.PreparedStatement;
import java.sql.ResultSet;
import java.sql.SQLException;
import java.sql.Statement;
import java.util.Vector;
import com.software.entity.MeetingRoomBean;
/**
 * 类名：MeetingRoomDao <br/>
 * 功能：封装对会议室信息表的数据库操作类. <br/>
 * 创建时间：2016-6-1 下午 10:17:24 <br/>
 * @author Administrator
 * @version
 * @since JDK 1.6
 */
public class MeetingRoomDao {
    /**
     * insertMeetingRoom：添加一条会议室信息. <br/>
     * @author    Administrator
     * @param meetingRoom 会议室实体类实例
     * @return 是否添加成功
     * @since JDK 1.6
     */
    public boolean insertMeetingRoom(MeetingRoomBean meetingRoom) {
        boolean isSuccess = false;
        Connection conn = null;
        PreparedStatement ps = null;
        try {
            String sql = "insert into t_meeting_room values(?,?)";
            conn = getConnection();
```

```java
            ps = conn.prepareStatement(sql);
            ps.setString(1, meetingRoom.getMeetingRoomID());
            ps.setString(2, meetingRoom.getMeetingRoomName());
            ps.executeUpdate();
            isSuccess = true;
        } catch (ClassNotFoundException e) {
            e.printStackTrace();
        } catch (SQLException e) {
            e.printStackTrace();
        } finally {
            try {
                if (ps != null)
                    ps.close();
                if (conn != null)
                    conn.close();
            } catch (SQLException e) {
                e.printStackTrace();
            }
        }
        return isSuccess;
    }
    /**
     * queryMeetingRoom:查询会议室信息. <br/>
     * @author      Administrator
     * @return 会议室信息查询结果
     * @since JDK 1.6
     */
    public Vector queryMeetingRoom(){
        Vector data=new Vector();
        Connection conn=null;
        Statement stm=null;
        ResultSet rs=null;
        String sql="select * from t_meeting_room";
        try {
            conn=getConnection();
            stm=conn.createStatement();
            rs=stm.executeQuery(sql);
            while(rs.next()){
                MeetingRoomBean meetingRoom=new MeetingRoomBean();
                meetingRoom.setMeetingRoomID(rs.getString("meeting_room_id"));
                meetingRoom.setMeetingRoomName(rs.getString("meeting_room_name"));
                data.add(meetingRoom);
            }
```

```java
            }catch (ClassNotFoundException e) {
                e.printStackTrace();
            } catch (SQLException e) {
                System.out.println(sql);
                e.printStackTrace();
            }finally{
                try {
                    if(rs!=null) rs.close();
                    if(stm!=null)stm.close();
                    if(conn!=null) conn.close();
                } catch (SQLException e) {
                    e.printStackTrace();
                }
            }
            return data;
    }
}
```

ReservationDao.java 代码

```java
package com.software.dao;
import static com.software.dao.DBUtil.getConnection;
import java.sql.Connection;
import java.sql.ResultSet;
import java.sql.SQLException;
import java.sql.Statement;
import java.util.Vector;
import javax.swing.table.DefaultTableModel;
import javax.swing.table.TableModel;
import com.software.entity.MeetingRoomBean;
import com.software.entity.ReservationBean;
/**
 * 类名：ReservationDao <br/>
 * 功能：封装会议室预订信息表数据库操作类 <br/>
 * 创建时间：2016-6-1 下午 11:05:33 <br/>
 * @author Administrator
 * @version
 * @since JDK 1.6
 */
public class ReservationDao {
    /**
     * getColumnNames：获取预订信息表格显示中文列名。<br/>
     * @author     Administrator
     * @return 预订信息表格显示中文列名
     * @since JDK 1.6
```

```java
 */
public static Vector<String> getColumnNames(){
    Vector<String> columnNames=new Vector<String>();
    columnNames.add(0,"编号");
    columnNames.add(1,"预订人");
    return columnNames;
}
/**
 * getEmpty：获取带中文列名预订信息空表数据模型. <br/>
 * @author    Administrator
 * @return 带中文列名预订信息空表数据模型
 * @since JDK 1.6
 */
public static TableModel getEmpty(){
    Vector data=new Vector();
    DefaultTableModel dmt=new DefaultTableModel(data,getColumnNames());
    return dmt;
}
/**
 * queryReservation：依据会议室编号和预订人姓名查询会议室预订信息. <br/>
 * @author    Administrator
 * @param meetingRoomID 会议室编号
 * @param reservationName 预订人姓名
 * @return 会议室预订信息查询结果
 * @since JDK 1.6
 */
public Vector queryReservation(String meetingRoomID,String reservationName){
    Vector data=new Vector();
    Connection conn=null;
    Statement stm=null;
    ResultSet rs=null;
    String sql="select * from t_reservation where (1=1) ";
    try {
        if(!(meetingRoomID==null||meetingRoomID.equals(""))){
            if(!(reservationName==null||reservationName.equals(""))){
                sql+=" and (meeting_room_id='"+meetingRoomID+"' and reservation_name='"+reservationName+"')";
            }else{
                sql+=" and (meeting_room_id='"+meetingRoomID+"')";
            }
        }
        conn=getConnection();
        stm=conn.createStatement();
```

```java
            rs=stm.executeQuery(sql);
            while(rs.next()){
                Vector row=new Vector();
                row.add(rs.getString("reservation_id"));
                row.add(rs.getString("reservation_name"));
                data.add(row);
            }
        }catch (ClassNotFoundException e) {
            e.printStackTrace();
        } catch (SQLException e) {
            System.out.println(sql);
            e.printStackTrace();
        }finally{
            try {
                if(rs!=null) rs.close();
                if(stm!=null)stm.close();
                if(conn!=null) conn.close();
            } catch (SQLException e) {
                e.printStackTrace();
            }
        }
        return data;
    }
    /**
     * querySingleStaff:依据预订编号查询会议室预订信息.<br/>
     * @author       Administrator
     * @param reservationId 预订编号
     * @return 会议室预订信息查询结果
     * @since JDK 1.6
     */
    public ReservationBean querySingleStaff(String reservationId){
        ReservationBean reservation=new ReservationBean();
        Connection conn=null;
        Statement stm=null;
        ResultSet rs=null;
        String sql="select a.meeting_room_name,b.* from t_meeting_room as a,t_reservation as b where a.meeting_room_id=b.meeting_room_id"
                +" and  b.reservation_id='"+reservationId+"'";
        try {
            conn=getConnection();
            stm=conn.createStatement();
            rs=stm.executeQuery(sql);
            while(rs.next()){
```

```java
                reservation.setReservationID(rs.getString("reservation_id"));
                reservation.setReservationName(rs.getString("reservation_name"));
                MeetingRoomBean meetingRoom = new MeetingRoomBean(rs.getString("meeting_room_id"), rs.getString("meeting_room_name"));
                reservation.setMeetingRoom(meetingRoom);
                reservation.setStartTime(rs.getString("start_time"));
                reservation.setEndTime(rs.getString("end_time"));
            }
        } catch (ClassNotFoundException e) {
            e.printStackTrace();
        } catch (SQLException e) {
            System.out.println(sql);
            e.printStackTrace();
        } finally {
            try {
                if(rs!=null) rs.close();
                if(stm!=null) stm.close();
                if(conn!=null) conn.close();
            } catch (SQLException e) {
                e.printStackTrace();
            }
        }
        return reservation;
    }
}
```

MeetingRoomBean.java 代码

```java
package com.software.entity;
/**
 * 类名:MeetingRoomBean <br/>
 * 功能:对应数据库中会议室信息表(t_meeting_room)实体类. <br/>
 * 创建时间:2016-6-1 下午 10:30:39 <br/>
 * @author Administrator
 * @version
 * @since JDK 1.6
 */
public class MeetingRoomBean {
    private String meetingRoomID;
    private String meetingRoomName;
    public MeetingRoomBean() {
    }
    public MeetingRoomBean(String meetingRoomID, String meetingRoomName) {
        this.meetingRoomID = meetingRoomID;
```

```java
        this.meetingRoomName = meetingRoomName;
    }
    public String getMeetingRoomID() {
        return meetingRoomID;
    }
    public void setMeetingRoomID(String meetingRoomID) {
        this.meetingRoomID = meetingRoomID;
    }
    public String getMeetingRoomName() {
        return meetingRoomName;
    }
    public void setMeetingRoomName(String meetingRoomName) {
        this.meetingRoomName = meetingRoomName;
    }
    /*
     * 重写 equals 方法,用于下拉列表
     */
    @Override
    public boolean equals(Object obj) {
        MeetingRoomBean meet=(MeetingRoomBean)obj;
        return meetingRoomID.equals(meet.getMeetingRoomID());
    }
    /*
     * 重写 toString 方法,用于下拉列表
     */
    @Override
    public String toString() {
        return meetingRoomName;
    }
}
```

ReservationBean.java 代码

```java
package com.software.entity;
/**
 * 类名:ReservationBean <br/>
 * 功能:对应数据库中会议预订信息表(t_reservation)实体类。<br/>
 * 创建时间:2016-6-1 下午 10:33:46 <br/>     *
 * @author Administrator
 * @version
 * @since JDK 1.6
 */
public class ReservationBean {
    private String reservationID;
```

```java
        private String reservationName;
        private MeetingRoomBean meetingRoom;
        private String   startTime;
        private String endTime;
        public String getReservationID() {
            return reservationID;
        }
        public void setReservationID(String reservationID) {
            this.reservationID = reservationID;
        }
        public String getReservationName() {
            return reservationName;
        }
        public void setReservationName(String reservationName) {
            this.reservationName = reservationName;
        }
        public MeetingRoomBean getMeetingRoom() {
            return meetingRoom;
        }
        public void setMeetingRoom(MeetingRoomBean meetingRoom) {
            this.meetingRoom = meetingRoom;
        }
        public String getStartTime() {
            return startTime;
        }
        public void setStartTime(String startTime) {
            this.startTime = startTime;
        }
        public String getEndTime() {
            return endTime;
        }
        public void setEndTime(String endTime) {
            this.endTime = endTime;
        }
}
```

AddMeetingRoomFrame.java 代码

```java
package com.software.ui;
import java.awt.Container;
import java.awt.event.ActionEvent;
import java.awt.event.ActionListener;
import java.awt.event.WindowAdapter;
import java.awt.event.WindowEvent;
```

```java
import javax.swing.JButton;
import javax.swing.JDialog;
import javax.swing.JFrame;
import javax.swing.JLabel;
import javax.swing.JOptionPane;
import javax.swing.JTextField;
import com.software.dao.MeetingRoomDao;
import com.software.entity.MeetingRoomBean;
/**
 * 类名：AddMeetingRoomFrame <br/>
 * 功能：添加会议室信息记录窗体类. <br/>
 * 创建时间：2016-6-1 下午 10:42:54 <br/>
 * @author Administrator
 * @version
 * @since JDK 1.6
 */
public class AddMeetingRoomFrame extends JDialog implements ActionListener{
    JLabel jbl_id=new JLabel("会议室编码：");
    JTextField jtf_id=new JTextField(30);
    JLabel jbl_name=new JLabel("会议室名称：");
    JTextField jtf_name=new JTextField(30);
    JButton jb_save=new JButton("保存");
    /**
     * 创建一个新的实例 AddMeetingRoomFrame,对界面控件布局.
     * @param parent 打开弹出窗体的上一级窗体实例
     */
    public AddMeetingRoomFrame(JFrame parent){
        super(parent,"会议室",true);
        Container container=getContentPane();
        container.setLayout(null);
        container.add(jbl_id);
        jbl_id.setBounds(50,50, 80, 30);
        container.add(jtf_id);
        jtf_id.setBounds(150,50,200,30);
        container.add(jbl_name);
        jbl_name.setBounds(50,100, 80, 30);
        container.add(jtf_name);
        jtf_name.setBounds(150,100,200,30);
        container.add(jb_save);
        jb_save.setBounds(150,150,80,30);
        jb_save.addActionListener(this);
        this.setSize(400, 300);
        this.setLocationRelativeTo(parent);
```

```java
            this.setVisible(true);
            this.addWindowListener(new WindowAdapter() {
                @Override
                public void windowClosing(WindowEvent e) {
                    dispose();
                }
            });
    }
    /**
     * saveMeetingRoom：添加会议室信息. <br/>
     * @author        Administrator
     * @return 添加是否成功
     * @since JDK 1.6
     */
    public boolean saveMeetingRoom(){
        MeetingRoomBean meetingRoom=new MeetingRoomBean();
        MeetingRoomDao meetingRoomDao=new MeetingRoomDao();
        meetingRoom.setMeetingRoomID(jtf_id.getText());
        meetingRoom.setMeetingRoomName(jtf_name.getText());
        return meetingRoomDao.insertMeetingRoom(meetingRoom);
    }
    /*
     * 按钮事件处理
     */
    @Override
    public void actionPerformed(ActionEvent e) {
        if(saveMeetingRoom()){
            JOptionPane.showMessageDialog(null,"保存成功");
            this.dispose();
        }
        else{
            JOptionPane.showMessageDialog(null,"保存不成功请检查数据");
        }
    }
}
```

ReservationDetailFrame.java 代码

```java
package com.software.ui;
import java.awt.Container;
import java.awt.event.ActionEvent;
import java.awt.event.ActionListener;
import java.awt.event.WindowAdapter;
import java.awt.event.WindowEvent;
```

```java
import javax.swing.JButton;
import javax.swing.JDialog;
import javax.swing.JFrame;
import javax.swing.JLabel;
import com.software.dao.ReservationDao;
import com.software.entity.ReservationBean;
/**
 * 类名：ReservationDetailFrame <br/>
 * 功能：显示会议室预订详细信息窗体类。<br/>
 * 创建时间：2016-6-1 下午 10:52:31 <br/>
 * @author Administrator
 * @version
 * @since JDK 1.6
 */
public class ReservationDetailFrame extends JDialog implements ActionListener {
    /**
     * 创建一个新的实例 ReservationDetailFrame,为界面控件布局.
     * @param parent 打开弹出窗体的上一级窗体实例
     * @param reservationID 预订编号
     */
    public ReservationDetailFrame(JDialog parent,String reservationID){
        super(parent,"会议室预订详细信息",true);
        ReservationDao reservationDao=new ReservationDao();
        ReservationBean reservation=reservationDao.querySingleStaff(reservationID);
        JLabel jbl_id=new JLabel("预订编号   "+reservation.getReservationID());
        JLabel jbl_name=new JLabel("预订人    "+reservation.getReservationName());
        JLabel jbl_startTime=new JLabel("开始时间   "+reservation.getStartTime());
        JLabel jbl_endTime=new JLabel("结束时间   "+reservation.getEndTime());
        JLabel jbl_meetingRoom=new JLabel("会议室 "+reservation.getMeetingRoom().getMeetingRoomName());

        JButton jb_close=new JButton("关闭");
        Container container=getContentPane();
        container.setLayout(null);
        container.add(jbl_id);
        jbl_id.setBounds(20, 20, 200, 30);
        container.add(jbl_name);
        jbl_name.setBounds(20, 60, 200, 30);
        container.add(jbl_startTime);
        jbl_startTime.setBounds(20, 100, 200, 30);
        container.add(jbl_endTime);
        jbl_endTime.setBounds(20, 140, 200, 30);
        container.add(jbl_meetingRoom);
        jbl_meetingRoom.setBounds(20, 180, 200, 30);
```

```java
            container.add(jb_close);
            jb_close.setBounds(120, 260, 80, 30);
            jb_close.addActionListener(this);
            this.setLocationRelativeTo(parent);
            this.setSize(300, 350);
            this.setVisible(true);
            this.addWindowListener(new WindowAdapter() {
                @Override
                public void windowClosing(WindowEvent e) {
                    dispose();
                }
            });
        }
        /*
         * 按钮事件处理
         */
        @Override
        public void actionPerformed(ActionEvent e) {
            this.dispose();
        }
    }
```

ReservationQueryFrame.java 代码

```java
package com.software.ui;
import java.awt.BorderLayout;
import java.awt.Container;
import java.awt.Dimension;
import java.awt.event.ActionEvent;
import java.awt.event.ActionListener;
import java.awt.event.WindowAdapter;
import java.awt.event.WindowEvent;
import java.util.Vector;
import javax.swing.BorderFactory;
import javax.swing.JButton;
import javax.swing.JComboBox;
import javax.swing.JDialog;
import javax.swing.JFrame;
import javax.swing.JLabel;
import javax.swing.JPanel;
import javax.swing.JScrollPane;
import javax.swing.JTable;
import javax.swing.JTextField;
import javax.swing.ListSelectionModel;
```

```java
import javax.swing.event.ListSelectionEvent;
import javax.swing.event.ListSelectionListener;
import com.software.dao.MeetingRoomDao;
import com.software.dao.MyTableModel;
import com.software.dao.ReservationDao;
import com.software.entity.MeetingRoomBean;
/**
 * 类名:ReservationQueryFrame <br/>
 * 功能:会议预订信息查询窗体类. <br/>
 * 创建时间:2016-6-1 下午10:59:25 <br/>
 * @author Administrator
 * @version
 * @since JDK 1.6
 */
public class ReservationQueryFrame extends JDialog implements ActionListener {
    JPanel jp_query = new JPanel();
    JLabel jbl_id=new JLabel("会议室");
    JComboBox jcb_id=null;
    JLabel jbl_name = new JLabel("预订人");
    JTextField jtf_name = new JTextField(30);
    JButton jb_query = new JButton("查询");
    JTable table = null;
    /**
     * 创建一个新的实例 ReservationQueryFrame,对界面控件进行布局.
     * @param parent 打开弹出窗体的上一级窗体实例
     */
    public ReservationQueryFrame(JFrame parent) {
        super(parent,"员工信息查询",true);
        jp_query.setLayout(null);
        MeetingRoomDao meetingRoomDao=new MeetingRoomDao();
        jcb_id=new JComboBox(meetingRoomDao.queryMeetingRoom());
        jp_query.add(jbl_id);
        jp_query.add(jcb_id);
        jbl_id.setBounds(20,20,80,30);
        jcb_id.setBounds(100,20,200,30);
        jp_query.add(jbl_name);
        jp_query.add(jtf_name);
        jbl_name.setBounds(20,60,80,30);
        jtf_name.setBounds(100,60,200,30);
        jp_query.add(jb_query);
        jb_query.setBounds(220,100,80,30);
        jb_query.addActionListener(this);
        table = new JTable(ReservationDao.getEmpty());
```

```java
            table.getSelectionModel().setSelectionMode(ListSelectionModel.SINGLE_SELECTION);
            table.getSelectionModel().addListSelectionListener(new SelectionListener());
            jp_query.setPreferredSize(new Dimension(350, 200));
            JScrollPane jsp = new JScrollPane(table);
            jsp.setBorder(BorderFactory.createTitledBorder("会议室预订信息"));
            Container container = getContentPane();
            container.add(jp_query, BorderLayout.NORTH);
            container.add(jsp);
            this.setSize(350, 500);
            this.setLocationRelativeTo(parent);
            this.setVisible(true);
            this.addWindowListener(new WindowAdapter() {
                @Override
                public void windowClosing(WindowEvent e) {
                    dispose();
                }
            });
        }
        /*
         * 按钮事件处理
         */
        @Override
        public void actionPerformed(ActionEvent e) {
            ReservationDao reservationDao = new ReservationDao();
            Vector data = reservationDao.queryReservation(((MeetingRoomBean)jcb_id.getSelectedItem()).getMeetingRoomID(), jtf_name.getText());
            MyTableModel myTableModel = new MyTableModel(data, ReservationDao.getColumnNames());
            table.setModel(myTableModel);
        }
        /**
         * 表格行选中事件监听类
         */
        class SelectionListener implements ListSelectionListener {
            public void valueChanged(ListSelectionEvent e) {
                int row = table.getSelectedRow();
                if(!e.getValueIsAdjusting() && row >= 0){
                    new ReservationDetailFrame(ReservationQueryFrame.this, (String)table.getValueAt(row, 0));
                    table.repaint();
                }
            }
        }
    }
```

MainFrame.java 代码
package com.software.ui；
import java.awt.Container；
import java.awt.event.ActionEvent；
import java.awt.event.ActionListener；
import javax.swing.JButton；
import javax.swing.JFrame；
/**
 * 类名：MainFrame
 * 功能：程序运行主窗体类。
 * 创建时间：2016－6－1 下午 10：47：26
 * @author Administrator
 * @version
 * @since JDK 1.6
 */
public class MainFrame extends JFrame implements ActionListener{
    JButton jb_add＝new JButton("会议室添加")；
    JButton   jb_query＝new JButton("预订查询")；
    /**
     * 创建一个新的实例 MainFrame，为界面控件布局.
     */
    public MainFrame(){
        Container c＝getContentPane()；
        c.setLayout(null)；
        c.add(jb_add)；
        jb_add.setBounds(50，100，200，30)；
        c.add(jb_query)；
        jb_query.setBounds(50,150，200，30)；
        jb_add.addActionListener(this)；
        jb_query.addActionListener(this)；
        this.setSize(300,300)；
        this.setTitle("通达办公自动化系统")；
        this.setLocationRelativeTo(null)；
        this.setVisible(true)；
        this.setDefaultCloseOperation(JFrame.EXIT_ON_CLOSE)；
    }
    /**
     * main：程序运行主函数。
     * @author   Administrator
     * @param args
     * @since JDK 1.6
     */

```java
    public static void main(String[] args) {
        new MainFrame();
    }
    /*
     * 按钮事件处理
     */
    @Override
    public void actionPerformed(ActionEvent e) {
        String cmd=e.getActionCommand();
        if(cmd.equals("会议室添加")){
            new AddMeetingRoomFrame(this);
        }else{
            new ReservationQueryFrame(this);
        }
    }
}
```

项目二十二

代码如下：
创建数据库代码
CREATE DATABASE oadb CHARACTER SET GBK；
USE oadb；

CREATE TABLE t_department(
    department_id VARCHAR(20) NOT NULL PRIMARY KEY,   ——部门编号
    department_name VARCHAR(20) NOT NULL   ——部门名称
);
INSERT INTO t_department
SELECT '001','办公室' UNION
SELECT '002','财务处' UNION
SELECT '003','人事处'；

CREATE TABLE t_staff(
    staff_id VARCHAR(20) NOT NULL PRIMARY KEY,   ——员工编号
    staff_name VARCHAR(20) NOT NULL,   ——姓名
    staff_sex VARCHAR(2) NOT NULL,   ——性别
    birthday datetime NOT NULL,   ——生日
    department_id VARCHAR(20) NOT NULL,   ——部门编号
    staff_password VARCHAR(20) NOT NULL   ——密码
);
INSERT INTO t_staff

SELECT '001','周明','男','1976-02-28','001','Test' UNION
SELECT '002','李青','女','1981-03-04','001','Test' UNION
SELECT '003','刘欣','男','1992-12-25','001','Test';

DBUtil.java 代码
package com.software.dao;
import java.sql.Connection;
import java.sql.DriverManager;
import java.sql.SQLException;
/**
 * 类名：DBUtil
 * 功能：自定义数据库工具类,封装通用数据库操作
 * 创建时间：2016-5-31 下午 4:51:22
 * @author Administrator
 * @version
 * @since JDK 1.6
 */
public class DBUtil {
    /**
     * getConnection:连接数据库返回,数据库连接对象.
     * @author   Administrator
     * @return 数据库连接对象
     * @throws ClassNotFoundException
     * @throws SQLException
     * @since JDK 1.6
     */
    public static Connection getConnection() throws ClassNotFoundException,SQLException{
        String url="jdbc:mysql://127.0.0.1:8306/oadb";
        String user="root";
        String pwd="1234";
        Class.forName("com.mysql.jdbc.Driver");
        Connection conn=DriverManager.getConnection(url, user, pwd);
        return conn;
    }
}

MyTableModel.java 代码
package com.software.dao;
import java.util.Vector;
import javax.swing.table.DefaultTableModel;
/**
 * 类名：MyTableModel

```
 * 功能：自定义 JTable 表格数据模型代码. <br/>
 * 创建时间：2016-5-31 下午 5:03:10 <br/>
 * @author Administrator
 * @version
 * @since JDK 1.6
 */
public class MyTableModel extends DefaultTableModel {
    /**
     * 创建一个新的实例 MyTableModel.
     * @param data 表格数据
     * @param columnNames 显示表格的列名
     */
    public MyTableModel(Vector data, Vector columnNames) {
        super(data, columnNames);
    }
    /**
     * 控制表格单元格是否可编辑.
     * @param r 行号
     * @param c 列号
     */
    public boolean isCellEditable(int r, int c) {
        return false;
    }
    /**
     * 获得单元格列类型.
     * @param c 列号
     * @return 列类型
     */
    public Class getColumnClass(int c) {
        return getValueAt(0, c).getClass();
    }
}
```

DepartmentDao.java 代码

```
package com.software.dao;
import static com.software.dao.DBUtil.getConnection;
import java.sql.Connection;
import java.sql.PreparedStatement;
import java.sql.ResultSet;
import java.sql.SQLException;
import java.sql.Statement;
import java.util.Vector;
import com.software.entity.DepartmentBean;
```

```java
/**
 * 类名：DepartmentDao <br/>
 * 功能：封装对部门信息表的数据库操作类. <br/>
 * 创建时间：2016-6-2 上午 9:17:47 <br/>
 * @author Administrator
 * @version
 * @since JDK 1.6
 */
public class DepartmentDao {
    /**
     * insertDepartment:添加部门信息记录. <br/>
     * @author       Administrator
     * @param department 部门信息实体类的实例
     * @return 是否添加部门信息数据成功
     * @since JDK 1.6
     */
    public boolean insertDepartment(DepartmentBean department) {
        boolean isSuccess = false;
        Connection conn = null;
        PreparedStatement ps = null;
        try {
            String sql = "insert into t_department values(?,?)";
            conn = getConnection();
            ps = conn.prepareStatement(sql);
            ps.setString(1, department.getDepartmentID());
            ps.setString(2, department.getDepartmentName());
            ps.executeUpdate();
            isSuccess = true;
        } catch (ClassNotFoundException e) {
            e.printStackTrace();
        } catch (SQLException e) {
            e.printStackTrace();
        } finally {
            try {
                if (ps != null)
                    ps.close();
                if (conn != null)
                    conn.close();
            } catch (SQLException e) {
                e.printStackTrace();
            }
        }
        return isSuccess;
```

```java
    }
    /**
     * queryDepartment:查询部门信息.<br/>
     * @author      Administrator
     * @return 返回部门信息查询结果
     * @since JDK 1.6
     */
    public Vector queryDepartment(){
        Vector data=new Vector();
        Connection conn=null;
        Statement stm=null;
        ResultSet rs=null;
        String sql="select * from t_department";
        try {
            conn=getConnection();
            stm=conn.createStatement();
            rs=stm.executeQuery(sql);
            while(rs.next()){
                DepartmentBean depart=new DepartmentBean();
                depart.setDepartmentID(rs.getString("department_id"));
                depart.setDepartmentName(rs.getString("department_name"));
                data.add(depart);
            }
        }catch (ClassNotFoundException e) {
            e.printStackTrace();
        } catch (SQLException e) {
            System.out.println(sql);
            e.printStackTrace();
        }finally{
            try {
                if(rs!=null) rs.close();
                if(stm!=null)stm.close();
                if(conn!=null) conn.close();
            } catch (SQLException e) {
                e.printStackTrace();
            }
        }
        return data;
    }
}
```

StaffDao.java 代码

```java
package com.software.dao;
import static com.software.dao.DBUtil.getConnection;
import java.sql.Connection;
import java.sql.ResultSet;
import java.sql.SQLException;
import java.sql.Statement;
import java.util.Vector;
import javax.swing.table.DefaultTableModel;
import javax.swing.table.TableModel;
import com.software.entity.DepartmentBean;
import com.software.entity.StaffBean;
/**
 * 类名：StaffDao <br/>
 * 功能：封装对员工信息表的数据库操作类. <br/>
 * 创建时间：2016-6-2 上午 9:26:01 <br/>
 * @author Administrator
 * @version
 * @since JDK 1.6
 */
public class StaffDao {
    /**
     * getColumnNames:获取员工信息显示表格中文列标题. <br/>
     * @author      Administrator
     * @return 员工信息显示表格中文列标题
     * @since JDK 1.6
     */
    public static Vector<String> getColumnNames(){
        Vector<String> columnNames=new Vector<String>();
        columnNames.add(0,"编号");
        columnNames.add(1,"员工姓名");
        return columnNames;
    }
    /**
     * getEmpty:获取带员工信息显示表格中文列标题的空表数据模型. <br/>
     * @author      Administrator
     * @return 带员工信息显示表格中文列标题的空表数据模型
     * @since JDK 1.6
     */
    public static TableModel getEmpty(){
        Vector data=new Vector();
        DefaultTableModel dmt=new DefaultTableModel(data,getColumnNames());
        return dmt;
    }
```

```java
/**
 * queryStaff:查询员工信息.<br/>
 * @author         Administrator
 * @param departmentID 部门编号
 * @param staffName 员工姓名
 * @return 员工信息查询结果
 * @since JDK 1.6
 */
public Vector queryStaff(String departmentID,String staffName){
    Vector data=new Vector();
    Connection conn=null;
    Statement stm=null;
    ResultSet rs=null;
    String sql="select * from t_staff where (1=1) ";
    try {
        if(! (departmentID==null||departmentID.equals(""))){
            if(! (staffName==null||staffName.equals(""))){
                sql+=" and (department_id='"+departmentID+"' and staff_name='"+staffName+"')";
            }else{
                sql+=" and (department_id='"+departmentID+"')";
            }
        }
        conn=getConnection();
        stm=conn.createStatement();
        rs=stm.executeQuery(sql);
        while(rs.next()){
            Vector row=new Vector();
            row.add(rs.getString("staff_id"));
            row.add(rs.getString("staff_name"));
            data.add(row);
        }
    }catch (ClassNotFoundException e) {
        e.printStackTrace();
    } catch (SQLException e) {
        System.out.println(sql);
        e.printStackTrace();
    }finally{
        try {
            if(rs!=null) rs.close();
            if(stm!=null)stm.close();
            if(conn!=null) conn.close();
        } catch (SQLException e) {
```

```java
                e.printStackTrace();
            }
        }
        return data;
}
/**
 * querySingleStaff:依据员工编号查询员工信息.<br/>
 * @author      Administrator
 * @param staffId 员工编号
 * @return 员工信息查询结果
 * @since JDK 1.6
 */
public StaffBean querySingleStaff(String staffId){
    StaffBean staff=new StaffBean();
    Connection conn=null;
    Statement stm=null;
    ResultSet rs=null;
    String sql="select a.*,b.department_name from t_staff as a,t_department as b where a.department_id=b.department_id"
            +" and   a.staff_id='"+staffId+"'";
    try {
        conn=getConnection();
        stm=conn.createStatement();
        rs=stm.executeQuery(sql);
        while(rs.next()){
            staff.setStaffID(rs.getString("staff_id"));
            staff.setStaffName(rs.getString("staff_name"));
            DepartmentBean department= new DepartmentBean(rs.getString("department_id"), rs.getString("department_name"));
            staff.setDepartment(department);
            staff.setStaffSex(rs.getString("staff_sex"));
            staff.setBirthday(rs.getString("birthday").substring(0,10));
            staff.setStaffPassword(rs.getString("staff_password"));
        }
    } catch (ClassNotFoundException e) {
        e.printStackTrace();
    } catch (SQLException e) {
        System.out.println(sql);
        e.printStackTrace();
    }finally{
        try {
            if(rs!=null) rs.close();
            if(stm!=null)stm.close();
```

```java
            if(conn! = null) conn.close();
        } catch (SQLException e) {
            e.printStackTrace();
        }
    }
    return staff;
}
```

DepartmentBean.java 代码

```java
package com.software.entity;
/**
 * 类名：DepartmentBean <br/>
 * 功能：对应数据库中部门信息表(t_department)的实体类。<br/>
 * 创建时间：2016-6-2 上午 9:30:08 <br/>
 * @author Administrator
 * @version
 * @since JDK 1.6
 */
public class DepartmentBean {
    private String departmentID;
    private String departmentName;
    public DepartmentBean() {
    }
    public DepartmentBean(String departmentID, String departmentName) {
        this.departmentID = departmentID;
        this.departmentName = departmentName;
    }
    public String getDepartmentID() {
        return departmentID;
    }
    public void setDepartmentID(String departmentID) {
        this.departmentID = departmentID;
    }
    public String getDepartmentName() {
        return departmentName;
    }
    public void setDepartmentName(String departmentName) {
        this.departmentName = departmentName;
    }
    /*
     * 重写 equals 方法，用于下拉列表
     */
```

```java
    @Override
    public boolean equals(Object obj) {
        DepartmentBean department=(DepartmentBean)obj;
        return department.getDepartmentID().equals(departmentID);
    }
    /*
     * 重写 toString,用于下拉列表
     */
    @Override
    public String toString() {
        return departmentName;
    }
}
```

StaffBean.java 代码

```java
package com.software.entity;
/**
 * 类名:StaffBean <br/>
 * 功能:对应数据库中员工信息表(t_staff)的实体类. <br/>
 * 创建时间:2016-6-2 上午 9:33:55 <br/>          *
 * @author Administrator
 * @version
 * @since JDK 1.6
 */
public class StaffBean {
    private String staffID;
    private String staffName;
    private DepartmentBean department;
    private String staffSex;
    private String birthday ;
    private String staffPassword;
    public String getStaffID() {
        return staffID;
    }
    public void setStaffID(String staffID) {
        this.staffID = staffID;
    }
    public String getStaffName() {
        return staffName;
    }
    public void setStaffName(String staffName) {
        this.staffName = staffName;
    }
```

```java
    public DepartmentBean getDepartment() {
        return department;
    }
    public void setDepartment(DepartmentBean department) {
        this.department = department;
    }
    public String getStaffSex() {
        return staffSex;
    }
    public void setStaffSex(String staffSex) {
        this.staffSex = staffSex;
    }
    public String getBirthday() {
        return birthday;
    }
    public void setBirthday(String birthday) {
        this.birthday = birthday;
    }
    public String getStaffPassword() {
        return staffPassword;
    }
    public void setStaffPassword(String staffPassword) {
        this.staffPassword = staffPassword;
    }

}
```

AddDepartmentFrame.java 代码

```java
package com.software.ui;
import java.awt.Container;
import java.awt.event.ActionEvent;
import java.awt.event.ActionListener;
import java.awt.event.WindowAdapter;
import java.awt.event.WindowEvent;
import javax.swing.JButton;
import javax.swing.JDialog;
import javax.swing.JFrame;
import javax.swing.JLabel;
import javax.swing.JOptionPane;
import javax.swing.JTextField;
import com.software.dao.DepartmentDao;
import com.software.entity.DepartmentBean;
/**
```

* 类名：AddDepartmentFrame
 * 功能：添加部门信息的窗体类.
 * 创建时间：2016-6-2 下午 2:53:08
 * @author Administrator
 * @version
 * @since JDK 1.6
 */
public class AddDepartmentFrame extends JDialog implements ActionListener{
    JLabel jbl_id=new JLabel("部门编码:");
    JTextField jtf_id=new JTextField(30);
    JLabel jbl_name=new JLabel("部门名称:");
    JTextField jtf_name=new JTextField(30);
    JButton jb_save=new JButton("保存");
    /**
     * 创建一个新的实例 AddDepartmentFrame,对界面控件进行布局.
     * @param parent 打开弹出窗体上一级窗体实体类
     */
    public AddDepartmentFrame(JFrame parent){
        super(parent,"部门",true);
        Container container=getContentPane();
        container.setLayout(null);
        container.add(jbl_id);
        jbl_id.setBounds(50,50, 80, 30);
        container.add(jtf_id);
        jtf_id.setBounds(150,50,200,30);
        container.add(jbl_name);
        jbl_name.setBounds(50,100, 80, 30);
        container.add(jtf_name);
        jtf_name.setBounds(150,100,200,30);
        container.add(jb_save);
        jb_save.setBounds(150,150,80,30);
        jb_save.addActionListener(this);
        this.setSize(400, 300);
        this.setLocationRelativeTo(parent);
        this.setVisible(true);
        this.addWindowListener(new WindowAdapter() {
            @Override
            public void windowClosing(WindowEvent e) {
                dispose();
            }
        });
    }
    /**

```java
 * saveDepartment:添加部门信息.<br/>
 * @author     Administrator
 * @return 返回添加是否成功
 * @since JDK 1.6
 */
public boolean saveDepartment(){
    DepartmentBean department=new DepartmentBean();
    DepartmentDao departmentDao=new DepartmentDao();
    department.setDepartmentID(jtf_id.getText());
    department.setDepartmentName(jtf_name.getText());
    return departmentDao.insertDepartment(department);
}
/*
 * 按钮事件处理
 */
@Override
public void actionPerformed(ActionEvent e) {
    if(saveDepartment()){
        JOptionPane.showMessageDialog(null,"保存成功");
        this.dispose();
    }
    else{
        JOptionPane.showMessageDialog(null,"保存不成功请检查数据");
    }
}
}
```

StaffDetailFrame.java 代码

```java
package com.software.ui;
import java.awt.Container;
import java.awt.event.ActionEvent;
import java.awt.event.ActionListener;
import java.awt.event.WindowAdapter;
import java.awt.event.WindowEvent;
import javax.swing.JButton;
import javax.swing.JDialog;
import javax.swing.JFrame;
import javax.swing.JLabel;
import import com.software.dao.StaffDao;
import com.software.entity.StaffBean;
/**
 * 类名:StaffDetailFrame<br/>
 * 功能:显示员工详细信息窗体类.<br/>
```

```java
 * 创建时间：2016-6-2 下午3:01:16 <br/>
 * @author Administrator
 * @version
 * @since JDK 1.6
 */
public class StaffDetailFrame extends JDialog implements ActionListener {
    /**
     * 创建一个新的实例 StaffDetailFrame,对界面控件进行布局.
     * @param parent 打开弹出窗体的上一级窗体实例
     * @param staffID 员工编号
     */
    public StaffDetailFrame(JDialog parent,String staffID){
        super(parent,"员工详细信息",true);
        StaffDao staffDao=new StaffDao();
        StaffBean staff=staffDao.querySingleStaff(staffID);
        JLabel jbl_id=new JLabel("员工编号　"+staff.getStaffID());
        JLabel jbl_name=new JLabel("名称　　"+staff.getStaffName());
        JLabel jbl_sex=new JLabel("性别　　"+staff.getStaffSex());
        JLabel jbl_birthday=new JLabel("生日　"+staff.getBirthday());
        JLabel jbl_department=new JLabel("部门"+staff.getDepartment().getDepartmentName());
        JLabel jbl_password=new JLabel("密码　"+staff.getStaffPassword());
        JButton jb_close=new JButton("关闭");
        Container container=getContentPane();
        container.setLayout(null);
        container.add(jbl_id);
        jbl_id.setBounds(20,20,200,30);
        container.add(jbl_name);
        jbl_name.setBounds(20,60,200,30);
        container.add(jbl_sex);
        jbl_sex.setBounds(20,100,200,30);
        container.add(jbl_birthday);
        jbl_birthday.setBounds(20,140,200,30);
        container.add(jbl_department);
        jbl_department.setBounds(20,180,200,30);
        container.add(jbl_password);
        jbl_password.setBounds(20,220,200,30);
        container.add(jb_close);
        jb_close.setBounds(120,260,80,30);
        jb_close.addActionListener(this);
        this.setLocationRelativeTo(parent);
        this.setSize(300,350);
        this.setVisible(true);
            this.addWindowListener(new WindowAdapter() {
```

```java
            @Override
            public void windowClosing(WindowEvent e) {
                dispose();
            }
        });
    }
    /*
     * 按钮事件处理
     */
    @Override
    public void actionPerformed(ActionEvent e) {
        this.dispose();
    }
}
```

StaffQueryFrame.java 代码

```java
package com.software.ui;
import java.awt.BorderLayout;
import java.awt.Container;
import java.awt.Dimension;
import java.awt.event.ActionEvent;
import java.awt.event.ActionListener;
import java.awt.event.WindowAdapter;
import java.awt.event.WindowEvent;
import java.util.Vector;
import javax.swing.BorderFactory;
import javax.swing.JButton;
import javax.swing.JComboBox;
import javax.swing.JDialog;
import javax.swing.JFrame;
import javax.swing.JLabel;
import javax.swing.JPanel;
import javax.swing.JScrollPane;
import javax.swing.JTable;
import javax.swing.JTextField;
import javax.swing.ListSelectionModel;
import javax.swing.event.ListSelectionEvent;
import javax.swing.event.ListSelectionListener;
import com.software.dao.DepartmentDao;import com.software.dao.MyTableModel;
import com.software.dao.StaffDao;
import com.software.entity.DepartmentBean;
/**
 * 类名：StaffQueryFrame <br/>
```

```
 * 功能：员工信息查询窗体类.<br/>
 * 创建时间：2016-6-2 下午3:05:48 <br/>
 * @author Administrator
 * @version
 * @since JDK 1.6
 */
public class StaffQueryFrame extends JDialog implements ActionListener {
    JPanel jp_query = new JPanel();
    JLabel jbl_id=new JLabel("部门");
    JComboBox jcb_id=null;
    JLabel jbl_name = new JLabel("员工姓名");
    JTextField jtf_name = new JTextField(30);
    JButton jb_query = new JButton("查询");
    JTable table = null;
    /**
     * 创建一个新的实例 StaffQueryFrame,对界面控件进行布局.
     * @param parent 打开弹出窗体上一级窗体的实例
     */
    public StaffQueryFrame(JFrame parent) {
        super(parent, "员工信息查询", true);
        jp_query.setLayout(null);
        DepartmentDao departmentDao=new DepartmentDao();
        jcb_id=new JComboBox(departmentDao.queryDepartment());
        jp_query.add(jbl_id);
        jp_query.add(jcb_id);
        jbl_id.setBounds(20, 20, 80, 30);
        jcb_id.setBounds(100, 20, 200, 30);
        jp_query.add(jbl_name);
        jp_query.add(jtf_name);
        jbl_name.setBounds(20, 60, 80, 30);
        jtf_name.setBounds(100, 60, 200, 30);
        jp_query.add(jb_query);
        jb_query.setBounds(220, 100, 80, 30);
        jb_query.addActionListener(this);
        table = new JTable(StaffDao.getEmpty());
        table.getSelectionModel().setSelectionMode(ListSelectionModel.SINGLE_SELECTION);
        table.getSelectionModel().addListSelectionListener(new SelectionListener());
        jp_query.setPreferredSize(new Dimension(350, 200));
        JScrollPane jsp = new JScrollPane(table);
        jsp.setBorder(BorderFactory.createTitledBorder("员工信息"));
        Container container = getContentPane();
        container.add(jp_query, BorderLayout.NORTH);
        container.add(jsp);
```

```java
        this.setSize(350, 500);
        this.setLocationRelativeTo(parent);
        this.setVisible(true);
        this.addWindowListener(new WindowAdapter() {
            @Override
            public void windowClosing(WindowEvent e) {
                dispose();
            }
        });
    }
    /**
     * 按钮事件处理
     */
    @Override
    public void actionPerformed(ActionEvent e) {
        StaffDao staffDao = new StaffDao();
        Vector data = staffDao.queryStaff(((DepartmentBean)jcb_id.getSelectedItem()).getDepartmentID(), jtf_name.getText());
        MyTableModel myTableModel = new MyTableModel(data, StaffDao.getColumnNames());
        table.setModel(myTableModel);
    }
    /**
     * 员工信息表格行选中事件监听类
     */
    class SelectionListener implements ListSelectionListener {
        public void valueChanged(ListSelectionEvent e) {
            int row = table.getSelectedRow();
            if(!e.getValueIsAdjusting()&&row>=0){
                new StaffDetailFrame(StaffQueryFrame.this,(String)table.getValueAt(row, 0));
                table.repaint();
            }
        }
    }
}
```

MainFrame.java 代码
```java
package com.software.ui;
import java.awt.Container;
import java.awt.event.ActionEvent;
import java.awt.event.ActionListener;
import javax.swing.JButton;
import javax.swing.JFrame;
/**
```

* 类名：MainFrame
 * 功能：程序运行主窗体类.
 * 创建时间：2016-6-2 上午 9:45:51
 * @author Administrator
 * @version
 * @since JDK 1.6
 */
public class MainFrame extends JFrame implements ActionListener{
    JButton jb_add=new JButton("部门添加");
    JButton jb_query=new JButton("员工信息查询");
    /**
     * 创建一个新的实例 MainFrame,对界面控件进行布局.
     */
    public MainFrame(){
        Container c=getContentPane();
        c.setLayout(null);
        c.add(jb_add);
        jb_add.setBounds(50,100,200,30);
        c.add(jb_query);
        jb_query.setBounds(50,150,200,30);
        jb_add.addActionListener(this);
        jb_query.addActionListener(this);
        this.setSize(300,300);
        this.setTitle("通达办公自动化系统");
        this.setLocationRelativeTo(null);
        this.setVisible(true);
        this.setDefaultCloseOperation(JFrame.EXIT_ON_CLOSE);
    }
    /**
     * main:程序运行主函数.
     * @author Administrator
     * @param args
     * @since JDK 1.6
     */
    public static void main(String[] args) {
        new MainFrame();
    }
    /*
     * 按钮事件处理
     */
    @Override
    public void actionPerformed(ActionEvent e) {
        String cmd=e.getActionCommand();

```
        if(cmd.equals("部门添加")){
            new AddDepartmentFrame(this);
        }else{
            new StaffQueryFrame(this);
        }
    }
}
```

项目二十三

代码如下：
CustomerDao.java 代码
```
package dao;
import java.sql.Connection;
import java.sql.DriverManager;
import java.sql.PreparedStatement;
import java.sql.ResultSet;
import java.sql.SQLException;
import java.util.ArrayList;
import java.util.Vector;
import entity.Customer;
/**
 * 类名：CustomerDao <br/>
 * 功能：封装客户信贷数据库操作类. <br/>
 * 创建时间：2016－6－2 下午 3:48:31 <br/>
 * @author Administrator
 * @version
 * @since JDK 1.6
 */
public class CustomerDao {
    Customer customer;
    Connection conn = null;
    ResultSet rs = null;
    PreparedStatement ps = null;
    /**
     * getConn:获取数据库连接. <br/>
     * @author     Administrator
     * @return 返回数据库连接对象
     * @since JDK 1.6
     */
    public Connection getConn() {
        try {
```

```java
            Class.forName("com.mysql.jdbc.Driver");
            conn = DriverManager.getConnection(
                    "jdbc:mysql://localhost:3306/BankCreditLoanDB2", "root",
                    "123456");
        } catch (ClassNotFoundException e) {

            e.printStackTrace();
        } catch (SQLException e) {

            e.printStackTrace();
        }
        return conn;
}
/**
 * queryByName:依据客户名称查询信贷信息.<br/>
 * @author      Administrator
 * @param name 客户名称
 * @return 信贷信息查询结果
 * @since JDK 1.6
 */
public Vector queryByName(String name) {
    Vector vector = new Vector();
    String sql = "select * from T_loan_detail where Cust_id='"+name+"'";
    conn = getConn();
    try {
        ps = conn.prepareStatement(sql);
        rs = ps.executeQuery();
        while (rs.next()) {
            Vector vc = new Vector();
            vc.add(rs.getString(1));
            vc.add(rs.getString(2));
            vc.add(rs.getString(3));
            vc.add(rs.getString(5));
            vector.add(vc);
        }
    } catch (SQLException e) {
        e.printStackTrace();
    }
    return vector;
}
/**
 * queryAllCustomer:查询所有客户信贷信息.<br/>
 * @author      Administrator
```

```java
 * @return 客户信贷信息查询结果
 * @since JDK 1.6
 */
public Vector queryAllCustomer() {
    Vector vector = new Vector();
    String sql = "select * from  T_Loan_detail ";
    conn = getConn();
    try {
        ps = conn.prepareStatement(sql);
        rs = ps.executeQuery();
        while (rs.next()) {
            Vector v = new Vector();
            v.add(rs.getString(1));
            v.add(rs.getInt(4));
            v.add(rs.getString(3));
            v.add(rs.getString(5));
            vector.add(v);
        }
    } catch (SQLException e) {
        e.printStackTrace();
    }
    return vector;
}
/**
 * queryCustomer:查询客户信息.<br/>
 * @author      Administrator
 * @return 客户信息查询结果
 * @since JDK 1.6
 */
public ArrayList queryCustomer() {
    ArrayList tcustomer = new ArrayList();
    String sql = "select * from  T_customer_info ";
    conn = getConn();
    try {
        ps = conn.prepareStatement(sql);
        rs = ps.executeQuery();
        while (rs.next()) {
            customer = new Customer();
            customer.setCust_id(rs.getString(1));
            customer.setCust_name(rs.getString(2));
            tcustomer.add(customer);
        }
    } catch (SQLException e) {
```

```
            e.printStackTrace();
        }
        return tcustomer;
    }
    /**
     * closeAll:关闭数据连接等.<br/>
     * @author      Administrator
     * @param conn 数据库连接对象
     * @param pstmt 语句对象
     * @param rs 结果集对象
     * @since JDK 1.6
     */
    public void closeAll(Connection conn, PreparedStatement pstmt, ResultSet rs) {
        if (rs != null) {
            try {
                rs.close();
            } catch (SQLException ex) {
                ex.printStackTrace();
            }
        }
        if (pstmt != null) {
            try {
                pstmt.close();
            } catch (SQLException ex) {
                ex.printStackTrace();
            }
        }
        if (conn != null) {
            try {
                conn.close();
            } catch (SQLException ex) {
                ex.printStackTrace();
            }
        }
    }
}
```

Customer.java 代码

```
package entity;
/**
 * 类名:Customer <br/>
 * 功能:对应数据库中客户信息表的实体类.<br/>
 * 创建时间:2016-6-2 下午3:49:29 <br/>          *
```

```java
 * @author Administrator
 * @version
 * @since JDK 1.6
 */
public class Customer {
    String Cust_id;
    String Cust_name;
    public Customer()
    {}
    public Customer(String id,String name)
    {
        Cust_id=id;
        Cust_name=name;
    }
    public String getCust_id() {
        return Cust_id;
    }
    public void setCust_id(String cust_id) {
        Cust_id = cust_id;
    }
    public String getCust_name() {
        return Cust_name;
    }
    public void setCust_name(String cust_name) {
        Cust_name = cust_name;
    }
}
```

LineTableModel.java 代码

```java
package model;
import java.util.Vector;
import javax.swing.table.DefaultTableModel;
/**
 * 类名：LineTableModel <br/>
 * 功能：自定义表格的数据模型. <br/>
 * 创建时间：2016-6-2 下午 3:50:23 <br/>
 * @author Administrator
 * @version
 * @since JDK 1.6
 */
public class LineTableModel extends DefaultTableModel {
    public LineTableModel(Vector v1,Vector v2){
        super(v1,v2);
```

        }
}

BankLoadFrame.java 代码
package ui;
import java.awt.BorderLayout;
import java.awt.EventQueue;
import java.awt.event.MouseEvent;
import java.util.ArrayList;
import java.util.Vector;
import javax.swing.JFrame;
import javax.swing.JPanel;
import javax.swing.border.EmptyBorder;
import javax.swing.event.ListSelectionEvent;
import dao.CustomerDao;
import entity.Customer;
import model.LineTableModel;
import javax.swing.DefaultListModel;
import javax.swing.GroupLayout;
import javax.swing.GroupLayout.Alignment;
import javax.swing.JList;
import javax.swing.JScrollPane;
import javax.swing.LayoutStyle.ComponentPlacement;
import javax.swing.JTable;
import javax.swing.JLabel;
import javax.swing.JTextField;
import javax.swing.JComboBox;
import javax.swing.JButton;
import java.awt.event.ActionListener;
import java.awt.event.ActionEvent;
/**
 * 类名：BankLoadFrame
 * 功能：银行信贷操作的窗体类.
 * 创建时间：2016-6-2 下午 3:51:56
 * @author Administrator
 * @version
 * @since JDK 1.6
 */
public class BankLoadFrame extends JFrame {
    private JPanel contentPane;
    CustomerDao cdao;
    DefaultListModel model;
    ArrayList<Customer> allCustomer;

```java
JList list;
private JPanel panel_1,panel_2;
private JScrollPane scrollPane_1;
private JTable table;
static   JTextField idTextField,moneyTextField;
JComboBox expireComboBox,transComboBox;
LineTableModel tablemodel;
JPanel panel;
JButton exitButton, addButton, delButton, saveButton;
boolean isUse = true;
JPanel panel_3;
/**
 * main:程序运行主函数.<br/>
 * @author     Administrator
 * @param args
 * @since JDK 1.6
 */
public static void main(String[] args) {
    EventQueue.invokeLater(new Runnable() {
        public void run() {
            try {
                BankLoadFrame frame = new BankLoadFrame();
                frame.setVisible(true);
            } catch (Exception e) {
                e.printStackTrace();
            }
        }
    });
}
/**
 * 创建一个新的实例 BankLoadFrame,对界面控件进行布局.
 */
public BankLoadFrame() {
    list = new JList();
    cdao = new CustomerDao();
    model = new DefaultListModel();
    queryCustomer();
    setDefaultCloseOperation(JFrame.EXIT_ON_CLOSE);
    setBounds(100, 100, 713, 300);
    contentPane = new JPanel();
    contentPane.setBorder(new EmptyBorder(5, 5, 5, 5));
    setContentPane(contentPane);
    panel = new JPanel();
```

```java
            panel_1 = new JPanel();
            panel_1.setBorder(javax.swing.BorderFactory.createTitledBorder("工程信息"));
            panel_2 = new JPanel();
            panel_3 = new JPanel();
            GroupLayout gl_contentPane = new GroupLayout(contentPane);
            gl_contentPane.setHorizontalGroup(
                gl_contentPane.createParallelGroup(Alignment.LEADING)
                    .addGroup(gl_contentPane.createSequentialGroup()
                        .addContainerGap()
                        .addComponent(panel, GroupLayout.PREFERRED_SIZE, 216, GroupLayout.PREFERRED_SIZE)
                        .addGroup(gl_contentPane.createParallelGroup(Alignment.TRAILING)
                            .addGroup(gl_contentPane.createSequentialGroup()
                                .addGap(51)
                                .addGroup(gl_contentPane.createParallelGroup(Alignment.LEADING)
                                    .addComponent(panel_1, GroupLayout.PREFERRED_SIZE, 383, GroupLayout.PREFERRED_SIZE)
                                    .addGroup(gl_contentPane.createSequentialGroup()
                                        .addGap(20)
                                        .addComponent(panel_3, GroupLayout.PREFERRED_SIZE, 343, GroupLayout.PREFERRED_SIZE)))
                                .addContainerGap(27, Short.MAX_VALUE))
                            .addGroup(gl_contentPane.createSequentialGroup()
                                .addPreferredGap(ComponentPlacement.RELATED)
                                .addComponent(panel_2, GroupLayout.PREFERRED_SIZE, 381, GroupLayout.PREFERRED_SIZE))))
            );
            gl_contentPane.setVerticalGroup(
                gl_contentPane.createParallelGroup(Alignment.LEADING)
                    .addGroup(gl_contentPane.createSequentialGroup()
                        .addGroup(gl_contentPane.createParallelGroup(Alignment.LEADING)
                            .addComponent(panel, GroupLayout.PREFERRED_SIZE, 190, GroupLayout.PREFERRED_SIZE)
                            .addGroup(gl_contentPane.createSequentialGroup()
                                .addComponent(panel_1, GroupLayout.PREFERRED_SIZE, 123, GroupLayout.PREFERRED_SIZE)
                                .addPreferredGap(ComponentPlacement.UNRELATED)
                                .addComponent(panel_2, GroupLayout.PREFERRED_SIZE, 72, GroupLayout.PREFERRED_SIZE)))
                        .addPreferredGap(ComponentPlacement.UNRELATED)
                        .addComponent(panel_3, GroupLayout.PREFERRED_SIZE, 30, GroupLayout.PREFERRED_SIZE)
                        .addContainerGap(GroupLayout.DEFAULT_SIZE, Short.MAX_VALUE))
```

```
        );
        addButton = new JButton("添加");
        addButton.addActionListener(new ActionListener() {
            public void actionPerformed(ActionEvent e) {
            }
        });
        saveButton = new JButton("保存");
        saveButton.addActionListener(new ActionListener() {
            public void actionPerformed(ActionEvent e) {
                isUse = false;
                addButton.setEnabled(true);
                saveButton.setEnabled(false);
                delButton.setEnabled(false);
            }
        });
        delButton = new JButton("删除");
        delButton.addActionListener(new ActionListener() {
            public void actionPerformed(ActionEvent e) {
            }
        });
        exitButton = new JButton("退出");
        exitButton.addActionListener(new ActionListener() {
            public void actionPerformed(ActionEvent e) {
            }
        });
        if (isUse) {
            saveButton.setEnabled(false);
            delButton.setEnabled(false);
        }
        GroupLayout gl_panel_3 = new GroupLayout(panel_3);
        gl_panel_3
            .setHorizontalGroup(
                gl_panel_3.createParallelGroup(Alignment.LEADING)
                    .addGroup(gl_panel_3.createSequentialGroup().addContainerGap().addComponent(addButton)
                        .addPreferredGap(ComponentPlacement.UNRELATED).addComponent(saveButton)
                        .addGap(18)
                        .addComponent(delButton, GroupLayout.PREFERRED_SIZE, 80,
                            GroupLayout.PREFERRED_SIZE)
                        .addPreferredGap(ComponentPlacement.RELATED).addComponent(exitButton)
```

```
                                .addContainerGap(12,Short.MAX_VALUE)));
        gl_panel_3.setVerticalGroup(gl_panel_3.createParallelGroup(Alignment.LEADING)
                .addGroup(gl_panel_3.createSequentialGroup()
                        .addGroup(gl_panel_3.createParallelGroup(Alignment.BASELINE).addCompo-
nent(addButton)
                                .addComponent(saveButton).addComponent(delButton).addComponent
(exitButton))
                        .addContainerGap(GroupLayout.DEFAULT_SIZE,Short.MAX_VALUE)));
        panel_3.setLayout(gl_panel_3);
        JLabel idLabel = new JLabel("货款编号");
        idTextField = new JTextField();
        idTextField.setEditable(false);
        idTextField.setColumns(10);
        JLabel transLabel = new JLabel("交易日期");
        transComboBox = new JComboBox();
        JLabel loanLabel = new JLabel("货款金额");
        moneyTextField = new JTextField();
        moneyTextField.setColumns(10);
        JLabel expireLabel = new JLabel("到期时间");
        expireComboBox = new JComboBox();
        GroupLayout gl_panel_2 = new GroupLayout(panel_2);
        gl_panel_2.setHorizontalGroup(
            gl_panel_2.createParallelGroup(Alignment.LEADING)
                .addGroup(gl_panel_2.createSequentialGroup()
                    .addContainerGap()
                    .addGroup(gl_panel_2.createParallelGroup(Alignment.LEADING)
                        .addGroup(gl_panel_2.createSequentialGroup()
                            .addComponent(idLabel)
                            .addPreferredGap(ComponentPlacement.RELATED)
                            .addComponent(idTextField,GroupLayout.PREFERRED_SIZE,Grou-
pLayout.DEFAULT_SIZE,GroupLayout.PREFERRED_SIZE))
                        .addGroup(gl_panel_2.createSequentialGroup()
                            .addComponent(loanLabel)
                            .addPreferredGap(ComponentPlacement.RELATED)
                            .addComponent(moneyTextField,GroupLayout.PREFERRED_SIZE,Grou-
pLayout.DEFAULT_SIZE,GroupLayout.PREFERRED_SIZE)))
                    .addPreferredGap(ComponentPlacement.RELATED,44,Short.MAX_VALUE)
                    .addGroup(gl_panel_2.createParallelGroup(Alignment.TRAILING)
                        .addGroup(gl_panel_2.createSequentialGroup()
                            .addComponent(transLabel)
                            .addGap(24))
                        .addGroup(gl_panel_2.createSequentialGroup()
                            .addComponent(expireLabel)
```

```java
                            .addGap(18)))
                        .addGroup(gl_panel_2.createParallelGroup(Alignment.LEADING, false)
                            .addComponent(expireComboBox, 0, GroupLayout.DEFAULT_SIZE, Short.MAX_VALUE)
                            .addComponent(transComboBox, 0, 101, Short.MAX_VALUE))
                        .addContainerGap(36, Short.MAX_VALUE))
        );
        gl_panel_2.setVerticalGroup(
            gl_panel_2.createParallelGroup(Alignment.LEADING)
                .addGroup(gl_panel_2.createSequentialGroup()
                    .addGroup(gl_panel_2.createParallelGroup(Alignment.BASELINE)
                        .addComponent(idLabel)
                        .addComponent(idTextField, GroupLayout.PREFERRED_SIZE, 17, GroupLayout.PREFERRED_SIZE)
                        .addComponent(transLabel)
                        .addComponent(transComboBox, GroupLayout.PREFERRED_SIZE, GroupLayout.DEFAULT_SIZE, GroupLayout.PREFERRED_SIZE))
                    .addPreferredGap(ComponentPlacement.UNRELATED)
                    .addGroup(gl_panel_2.createParallelGroup(Alignment.LEADING)
                        .addGroup(gl_panel_2.createParallelGroup(Alignment.BASELINE)
                            .addComponent(moneyTextField, GroupLayout.PREFERRED_SIZE, GroupLayout.DEFAULT_SIZE, GroupLayout.PREFERRED_SIZE)
                            .addComponent(loanLabel))
                        .addGroup(gl_panel_2.createParallelGroup(Alignment.BASELINE)
                            .addComponent(expireComboBox, GroupLayout.PREFERRED_SIZE, GroupLayout.DEFAULT_SIZE, GroupLayout.PREFERRED_SIZE)
                            .addComponent(expireLabel)))
                    .addGap(20))
        );
        panel_2.setLayout(gl_panel_2);
        scrollPane_1 = new JScrollPane();
        GroupLayout gl_panel_1 = new GroupLayout(panel_1);
        gl_panel_1.setHorizontalGroup(gl_panel_1.createParallelGroup(Alignment.LEADING)
            .addGroup(gl_panel_1.createSequentialGroup().addContainerGap()
                .addComponent(scrollPane_1, GroupLayout.PREFERRED_SIZE, 380, GroupLayout.PREFERRED_SIZE)
                .addContainerGap(GroupLayout.DEFAULT_SIZE, Short.MAX_VALUE)));
        gl_panel_1.setVerticalGroup(gl_panel_1.createParallelGroup(Alignment.LEADING).addComponent(scrollPane_1,
            GroupLayout.DEFAULT_SIZE, 99, Short.MAX_VALUE));
        panel_1.setLayout(gl_panel_1);
        Vector v = cdao.queryAllCustomer();
        table = new JTable();
```

```java
        table.addMouseListener(new java.awt.event.MouseAdapter() {
            public void mousePressed(java.awt.event.MouseEvent evt) {
                tableMousePressed(evt);
            }
            private void tableMousePressed(MouseEvent evt) {
            }
            public void mouseReleased(java.awt.event.MouseEvent evt) {
                tableMouseReleased(evt);
            }
            private void tableMouseReleased(MouseEvent evt) {
                addButton.setEnabled(false);
                saveButton.setEnabled(true);
                delButton.setEnabled(true);
                idTextField.setText((String) table.getValueAt(table.getSelectedRow(), 0));
                moneyTextField.setText((String) table.getValueAt(table.getSelectedRow(), 1));
                expireComboBox.addItem(table.getValueAt(table.getSelectedRow(), 2));
                transComboBox.addItem(table.getValueAt(table.getSelectedRow(), 3));
            }
        });
        JScrollPane scrollPane = new JScrollPane();
        GroupLayout gl_panel = new GroupLayout(panel);
        gl_panel.setHorizontalGroup(
                gl_panel.createParallelGroup(Alignment.LEADING)
                    .addGroup(gl_panel.createSequentialGroup().addContainerGap()
                        .addComponent(scrollPane, GroupLayout.DEFAULT_SIZE, 131, Short.MAX_VALUE)
                        .addContainerGap()));
        gl_panel.setVerticalGroup(gl_panel.createParallelGroup(Alignment.LEADING)
                    .addGroup(gl_panel.createSequentialGroup().addContainerGap()
                        .addComponent(scrollPane, GroupLayout.PREFERRED_SIZE, 161, GroupLayout.PREFERRED_SIZE)
                        .addContainerGap(19, Short.MAX_VALUE)));
        list.addListSelectionListener(new javax.swing.event.ListSelectionListener() {
            @Override
            public void valueChanged(ListSelectionEvent e) {
                isValuedChanged(e);
            }
        });
        scrollPane_1.setViewportView(table);
        scrollPane.setViewportView(list);
        panel.setLayout(gl_panel);
        contentPane.setLayout(gl_contentPane);
    }
```

```java
/**
 * getColumens:获取表格中文列标题. <br/>
 * @author      Administrator
 * @return 表格中文列标题
 * @since JDK 1.6
 */
public Vector getColumens() {
    Vector vc = new Vector();
    vc.add("贷款编号");
    vc.add("贷款数量");
    vc.add("交易日期");
    vc.add("到期时间");
    return vc;
}
/**
 * isValuedChanged:判断列表值是否发生改变. <br/>
 * @author      Administrator
 * @param e
 * @since JDK 1.6
 */
private void isValuedChanged(ListSelectionEvent e) {
    if (this.list.getSelectedValue() != null) {
        String name = (String) this.list.getSelectedValue();
        Vector v = cdao.queryByName(name);
        tablemodel = new LineTableModel(v, getColumens());
        table.setModel(tablemodel);
    }
}
/**
 * queryCustomer:查询所有的客户信息. <br/>
 * @author      Administrator
 * @since JDK 1.6
 */
private void queryCustomer() {
    model.clear();
    allCustomer = cdao.queryCustomer();
    if (!allCustomer.isEmpty()) {
        for (int i = 0; i < allCustomer.size(); i++)
            model.addElement(allCustomer.get(i).getCust_name());
        System.out.println(allCustomer.get(2).getCust_name());
    }
    this.list.setModel(model);
}
```

}

项目二十四

代码如下：

CustomerDao.java 代码

```java
package dao;
import java.sql.Connection;
import java.sql.DriverManager;
import java.sql.PreparedStatement;
import java.sql.ResultSet;
import java.sql.SQLException;
import java.util.ArrayList;
import java.util.Vector;
import entity.Customer;
/**
 * 类名：CustomerDao <br/>
 * 功能：封装客户信贷数据库操作类. <br/>
 * 创建时间：2016-6-2 下午 3:48:31 <br/>
 * @author Administrator
 * @version
 * @since JDK 1.6
 */
public class CustomerDao {
    Customer customer;
    Connection conn = null;
    ResultSet rs = null;
    PreparedStatement ps = null;
    /**
     * getConn:获取数据库连接. <br/>
     * @author Administrator
     * @return 返回数据库连接对象
     * @since JDK 1.6
     */
    public Connection getConn() {
        try {
            Class.forName("com.mysql.jdbc.Driver");
            conn = DriverManager.getConnection("jdbc:mysql://localhost:3306/BankCreditLoanDB", "root", "123456");
        } catch (ClassNotFoundException e) {
            e.printStackTrace();
```

```java
        } catch (SQLException e) {

            e.printStackTrace();
        }
        return conn;
}
/**
 * queryAllCustomer:查询所有的客户信息.<br/>
 * @author         Administrator
 * @return 客户信息查询结果
 * @since JDK 1.6
 */
public Vector queryAllCustomer() {
    Vector vector = new Vector();
    String sql = "select * from  T_customer_info ";
    conn = getConn();
    try {
        ps = conn.prepareStatement(sql);
        rs = ps.executeQuery();
        while (rs.next()) {
            Vector v = new Vector();
            v.add(rs.getString(1));
            v.add(rs.getString(2));
            v.add(rs.getString(3));
            v.add(rs.getString(4));
            v.add(rs.getString(5));
            vector.add(v);
        }

    } catch (SQLException e) {
        e.printStackTrace();
    }
    return vector;
}
/**
 * insertCustomer:添加客户信息.<br/>
 * @author         Administrator
 * @param cs 客户实体类实例
 * @since JDK 1.6
 */
public void insertCustomer(Customer cs) {
    String sql = "insert into T_customer_info values('" + cs.getCust_id() + "','" + cs.getCust_name() + "','"
```

```java
                    + cs.getLegal_name() + "','" + cs.getReg_address() + "','" + cs.getPost_code() + "')";
            conn = getConn();
            try {
                ps = conn.prepareStatement(sql);
                int i = ps.executeUpdate();
                if (i > 0)
                    System.out.println("hello");
            } catch (SQLException e) {
                e.printStackTrace();
            }
        }
        /**
         * closeAll:关闭数据连接等. <br/>
         * @author      Administrator
         * @param conn 数据库连接对象
         * @param pstmt 语句对象
         * @param rs 结果集对象
         * @since JDK 1.6
         */
        public void closeAll(Connection conn, PreparedStatement pstmt, ResultSet rs) {
            if (rs != null) {
                try {
                    rs.close();
                } catch (SQLException ex) {
                    ex.printStackTrace();
                }
            }
            if (pstmt != null) {
                try {
                    pstmt.close();
                } catch (SQLException ex) {
                    ex.printStackTrace();
                }
            }
            if (conn != null) {
                try {
                    conn.close();
                } catch (SQLException ex) {
                    ex.printStackTrace();
                }
            }
        }
```

Customer.java 代码

```java
package entity;
/**
 * 类名：Customer <br/>
 * 功能：对数据库中客户信息表的实体类. <br/>
 * 创建时间：2016-6-2 下午5:14:23 <br/>
 * @author Administrator
 * @version
 * @since JDK 1.6
 */
public class Customer {
    String Cust_id;
    String Cust_name;
    String Legal_name;
    String Reg_address;
    String Post_code;
    public Customer() {
    }
    public Customer(String Cust_id, String Cust_name, String Legal_name,
            String Reg_address, String Post_code) {
        this.Legal_name = Legal_name;
        this.Reg_address = Reg_address;
        this.Cust_name = Cust_name;
        this.Cust_id = Cust_id;
        this.Post_code = Post_code;
    }
    public String getCust_id() {
        return Cust_id;
    }
    public void setCust_id(String cust_id) {
        Cust_id = cust_id;
    }
    public String getCust_name() {
        return Cust_name;
    }
    public void setCust_name(String cust_name) {
        Cust_name = cust_name;
    }
    public String getLegal_name() {
        return Legal_name;
    }
```

```java
        public void setLegal_name(String legal_name) {
            Legal_name = legal_name;
        }
        public String getReg_address() {
            return Reg_address;
        }
        public void setReg_address(String reg_address) {
            Reg_address = reg_address;
        }
        public String getPost_code() {
            return Post_code;
        }
        public void setPost_code(String post_code) {
            Post_code = post_code;
        }
}
```

LineTableModel.java 代码

```java
package model;
import java.util.Vector;
import javax.swing.table.DefaultTableModel;
/**
 * 类名：LineTableModel <br/>
 * 功能：自定义表格的数据模型。<br/>
 * 创建时间：2016-6-2 下午3:50:23 <br/>
 * @author Administrator
 * @version
 * @since JDK 1.6
 */
public class LineTableModel extends DefaultTableModel {
    public LineTableModel(Vector v1,Vector v2){
        super(v1,v2);
    }
}
```

BankFrame.java 代码

```java
package ui;
import java.awt.EventQueue;
import java.awt.event.ActionEvent;
import java.awt.event.ActionListener;
import java.text.SimpleDateFormat;
import java.util.Date;
import java.util.Random;
```

```java
import java.util.Vector;
import javax.swing.GroupLayout;
import javax.swing.GroupLayout.Alignment;
import javax.swing.JButton;
import javax.swing.JFrame;
import javax.swing.JLabel;
import javax.swing.JPanel;
import javax.swing.JScrollPane;
import javax.swing.JTable;
import javax.swing.JTextField;
import javax.swing.LayoutStyle.ComponentPlacement;
import javax.swing.SwingConstants;
import javax.swing.border.EmptyBorder;
import dao.CustomerDao;
import entity.Customer;
import model.LineTableModel;
/**
 * 类名：BankFrame <br/>
 * 功能：银行客户信息管理窗体类. <br/>
 * 创建时间：2016-6-2 下午 4:07:12 <br/>
 * @author Administrator
 * @version
 * @since JDK 1.6
 */
public class BankFrame extends JFrame {
    private boolean isEnable = true;
    private JPanel contentPane;
    private JTextField bhtextField, daibiaoTextField, postCodeTextField, nameTextField, regTextField;
    private JTable table;
    LineTableModel ltable;
    JButton delButton, exitButton, modifyButton, saveButton, addButton, cancelButton, queryButton;
    Vector vc;
    CustomerDao cdao = new CustomerDao();
    /**
     * main:程序运行主函数. <br/>
     * @author      Administrator
     * @param args
     * @since JDK 1.6
     */
    public static void main(String[] args) {
        EventQueue.invokeLater(new Runnable() {
            public void run() {
                try {
```

```java
                    BankFrame frame = new BankFrame();
                    frame.setVisible(true);
                } catch (Exception e) {
                    e.printStackTrace();
                }
            }
        });
    }
    /**
     * 创建一个新的实例 BankFrame,对界面控件进行布局.
     */
    public BankFrame() {
        setDefaultCloseOperation(JFrame.EXIT_ON_CLOSE);
        setBounds(100, 100, 656, 413);
        contentPane = new JPanel();
        contentPane.setBorder(new EmptyBorder(5, 5, 5, 5));
        setContentPane(contentPane);
        JPanel panel = new JPanel();
        JPanel panel_1 = new JPanel();
        JPanel panel_2 = new JPanel();
        JScrollPane scrollPane = new JScrollPane();
        table = new JTable();
        vc = cdao.queryAllCustomer();
        ltable = new LineTableModel(null, this.getColumn());
        table.setModel(ltable);
        scrollPane.setViewportView(table);
        GroupLayout gl_contentPane = new GroupLayout(contentPane);
        gl_contentPane.setHorizontalGroup(gl_contentPane.createParallelGroup(Alignment.LEADING)
            .addGroup(gl_contentPane.createSequentialGroup()
                .addGroup(gl_contentPane.createParallelGroup(Alignment.LEADING)
                    .addComponent(panel, GroupLayout.PREFERRED_SIZE, 540, GroupLayout.PREFERRED_SIZE)
                    .addGroup(gl_contentPane.createSequentialGroup().addGap(22)
                        .addComponent(scrollPane, GroupLayout.PREFERRED_SIZE, 472,
                            GroupLayout.PREFERRED_SIZE)
                        .addPreferredGap(ComponentPlacement.RELATED).addComponent(panel_2,
                            GroupLayout.PREFERRED_SIZE, 492, GroupLayout.PREFERRED_SIZE))
                    .addGroup(gl_contentPane.createSequentialGroup().addContainerGap().addComponent(panel_1,
                        GroupLayout.PREFERRED_SIZE, 538, GroupLayout.PREFERRED_
```

```
SIZE)))
                              .addContainerGap(82, Short.MAX_VALUE)));
              gl_contentPane.setVerticalGroup(gl_contentPane.createParallelGroup(Alignment.LEADING).ad-
dGroup(gl_contentPane
                       .createSequentialGroup().addContainerGap()
                       .addComponent(panel, GroupLayout.PREFERRED_SIZE, 45, GroupLayout.PREF-
ERRED_SIZE).addGap(18)
                       .addComponent(panel_1, GroupLayout.PREFERRED_SIZE, 105, GroupLayout.PREF-
ERRED_SIZE)
                       .addGroup(gl_contentPane.createParallelGroup(Alignment.LEADING)
                            .addGroup(gl_contentPane.createSequentialGroup()
                                .addPreferredGap(ComponentPlacement.RELATED, 65, Short.MAX_
VALUE)
                                .addComponent(panel_2, GroupLayout.PREFERRED_SIZE, 122,
GroupLayout.PREFERRED_SIZE))
                            .addGroup(gl_contentPane.createSequentialGroup().addGap(18)
                                .addComponent(scrollPane, GroupLayout.PREFERRED_SIZE, 122,
GroupLayout.PREFERRED_SIZE)
                                .addContainerGap())));
              GroupLayout gl_panel_2 = new GroupLayout(panel_2);
              gl_panel_2
                       .setHorizontalGroup(gl_panel_2.createParallelGroup(Alignment.LEADING).addGap(0,
492, Short.MAX_VALUE));
              gl_panel_2.setVerticalGroup(gl_panel_2.createParallelGroup(Alignment.LEADING).addGap(0,
122, Short.MAX_VALUE));
              panel_2.setLayout(gl_panel_2);
              JLabel khbhLabel = new JLabel("客户编号");
              bhtextField = new JTextField();
              bhtextField.setColumns(10);
              JLabel lblNewLabel_1 = new JLabel("(由系统自动生成)");
              JLabel daibiaoLabel = new JLabel("法人代表");
              daibiaoTextField = new JTextField();
              daibiaoTextField.setColumns(10);
              JLabel nameLabel = new JLabel("客户名称");
              postCodeTextField = new JTextField();
              postCodeTextField.setColumns(10);
              JLabel postLabel = new JLabel("邮政编码");
              nameTextField = new JTextField();
              nameTextField.setColumns(10);
              JLabel regLabel = new JLabel("注册地址");
              regTextField = new JTextField();
              regTextField.setColumns(10);
              GroupLayout gl_panel_1 = new GroupLayout(panel_1);
```

```java
gl_panel_1.setHorizontalGroup(gl_panel_1.createParallelGroup(Alignment.LEADING)
    .addGroup(gl_panel_1.createSequentialGroup().addContainerGap()
        .addGroup(gl_panel_1.createParallelGroup(Alignment.LEADING)
            .addGroup(gl_panel_1.createSequentialGroup().addComponent(khbhLabel)
                .addPreferredGap(ComponentPlacement.UNRELATED)
                .addComponent(bhtextField, GroupLayout.PREFERRED_SIZE, GroupLayout.DEFAULT_SIZE, GroupLayout.PREFERRED_SIZE)
                .addGap(18).addComponent(lblNewLabel_1))
            .addGroup(gl_panel_1.createSequentialGroup()
                .addGroup(gl_panel_1.createParallelGroup(Alignment.LEADING)
                    .addGroup(gl_panel_1.createSequentialGroup().addGap(6)
                        .addComponent(daibiaoLabel).addPreferredGap(ComponentPlacement.RELATED)
                        .addComponent(daibiaoTextField, GroupLayout.PREFERRED_SIZE, GroupLayout.DEFAULT_SIZE, GroupLayout.PREFERRED_SIZE)
                        .addGap(27).addComponent(postLabel))
                    .addGroup(gl_panel_1.createSequentialGroup().addComponent(nameLabel)
                        .addPreferredGap(ComponentPlacement.UNRELATED)
                        .addComponent(postCodeTextField, GroupLayout.PREFERRED_SIZE, GroupLayout.DEFAULT_SIZE, GroupLayout.PREFERRED_SIZE)
                        .addGap(18).addComponent(regLabel)))
                .addGap(18)
                .addGroup(gl_panel_1.createParallelGroup(Alignment.LEADING)
                    .addGroup(gl_panel_1.createSequentialGroup().addGap(10).addComponent(
                        regTextField, GroupLayout.PREFERRED_SIZE, GroupLayout.DEFAULT_SIZE, GroupLayout.PREFERRED_SIZE))
                    .addComponent(nameTextField, GroupLayout.PREFERRED_SIZE, GroupLayout.DEFAULT_SIZE, GroupLayout.PREFERRED_SIZE))))
        .addContainerGap(326, Short.MAX_VALUE)));
gl_panel_1.setVerticalGroup(gl_panel_1.createParallelGroup(Alignment.LEADING)
    .addGroup(gl_panel_1.createSequentialGroup().addContainerGap()
```

```
                    .addGroup(gl_panel_1.createParallelGroup(Alignment.BASELINE).addCompo-
nent(khbhLabel)
                    .addComponent(bhtextField, GroupLayout.PREFERRED_SIZE, Grou-
pLayout.DEFAULT_SIZE,
                        GroupLayout.PREFERRED_SIZE)
                    .addComponent(lblNewLabel_1))
                .addPreferredGap(ComponentPlacement.UNRELATED)
                .addGroup(gl_panel_1.createParallelGroup(Alignment.LEADING)
                    .addGroup(gl_panel_1.createParallelGroup(Alignment.BASELINE)
                        .addComponent(daibiaoTextField, GroupLayout.PREFERRED_SIZE,
GroupLayout.DEFAULT_SIZE,
                            GroupLayout.PREFERRED_SIZE)
                        .addComponent(postLabel).addComponent(nameTextField, GroupLay-
out.PREFERRED_SIZE,
                            GroupLayout.DEFAULT_SIZE, GroupLayout.PREFERRED
_SIZE))
                        .addComponent(daibiaoLabel))
                .addPreferredGap(ComponentPlacement.RELATED)
                .addGroup(gl_panel_1.createParallelGroup(Alignment.LEADING)
                    .addGroup(gl_panel_1.createParallelGroup(Alignment.BASELINE).addCompo-
nent(nameLabel)
                        .addComponent(postCodeTextField, GroupLayout.PREFERRED_
SIZE, GroupLayout.DEFAULT_SIZE,
                            GroupLayout.PREFERRED_SIZE))
                    .addGroup(gl_panel_1.createParallelGroup(Alignment.BASELINE).addCompo-
nent(regLabel)
                        .addComponent(regTextField, GroupLayout.PREFERRED_SIZE,
GroupLayout.DEFAULT_SIZE,
                            GroupLayout.PREFERRED_SIZE)))
                .addContainerGap(16, Short.MAX_VALUE)));
        panel_1.setLayout(gl_panel_1);
        addButton = new JButton("添加");
        addButton.addActionListener(new ActionListener() {
            public void actionPerformed(ActionEvent e) {
                saveButton.setEnabled(true);
                cancelButton.setEnabled(true);
                Date now = new Date();
                SimpleDateFormat df = new SimpleDateFormat("yyyyMMdd");// 设置日期格式
                String bh = df.format(now);
                Random r = new Random();
                int x = r.nextInt(10);
                int y = r.nextInt(10);
                String id = bh + x + y;
```

```java
                    bhtextField.setText(id);
                }
        });
        modifyButton = new JButton("修改");
        saveButton = new JButton("保存");
        saveButton.addActionListener(new ActionListener() {
            public void actionPerformed(ActionEvent e) {
                saveButton.setEnabled(false);
                cancelButton.setEnabled(false);
                String daibiao = daibiaoTextField.getText();
                String customername = nameTextField.getText();
                String postcode = postCodeTextField.getText();
                String address = regTextField.getText();
                String id = bhtextField.getText();
                Customer cs = new Customer(id, customername, daibiao, address, postcode);
                cdao.insertCustomer(cs);
            }
        });
        saveButton.setVerticalAlignment(SwingConstants.BOTTOM);
        cancelButton = new JButton("取消");
        queryButton = new JButton("查询");
        queryButton.addActionListener(new ActionListener() {
            public void actionPerformed(ActionEvent e) {
                ltable = new LineTableModel(vc, getColumn());
                table.setModel(ltable);
            }
        });
        delButton = new JButton("删除");
        exitButton = new JButton("退出");
        GroupLayout gl_panel = new GroupLayout(panel);
        gl_panel.setHorizontalGroup(gl_panel.createParallelGroup(Alignment.LEADING)
                .addGroup(gl_panel.createSequentialGroup().addComponent(addButton).addGap(29).addComponent(modifyButton)
                        .addGap(18).addComponent(saveButton).addGap(18).addComponent(cancelButton).addGap(18)
                        .addComponent(queryButton).addPreferredGap(ComponentPlacement.RELATED).addComponent(delButton)
                        .addPreferredGap(ComponentPlacement.UNRELATED).addComponent(exitButton)
                        .addContainerGap(164, Short.MAX_VALUE)));
        gl_panel.setVerticalGroup(gl_panel.createParallelGroup(Alignment.LEADING).addGroup(Alignment.TRAILING,
                gl_panel.createSequentialGroup().addContainerGap(22, Short.MAX_VALUE)
```

```
                        .addGroup(gl_panel.createParallelGroup(Alignment.BASELINE).addComponent
(addButton)
                                .addComponent(modifyButton).addComponent(saveButton).addCom-
ponent(cancelButton)
                                .addComponent(queryButton).addComponent(delButton).addCompo-
nent(exitButton))
                            .addContainerGap()));
        panel.setLayout(gl_panel);
        contentPane.setLayout(gl_contentPane);
        if(isEnable)
        {
            modifyButton.setEnabled(false);
            saveButton.setEnabled(false);
            delButton.setEnabled(false);
            cancelButton.setEnabled(false);
        }
    }
    /**
     * getColumn:获取表格的中文列标题.<br/>
     * @author      Administrator
     * @return 表格的中文列标题.
     * @since JDK 1.6
     */
    public Vector getColumn() {
        Vector vec = new Vector();
        vec.add("客户编号");
        vec.add("客户名称");
        vec.add("法人代表");
        vec.add("注册地址");
        vec.add("邮政编码");
        return vec;
    }
}
```